First-Principles Prediction of Structures and Properties in Crystals

First-Principles Prediction of Structures and Properties in Crystals

Special Issue Editors

Andreas Hermann
Dominik Kurzydłowski

MDPI • Basel • Beijing • Wuhan • Barcelona • Belgrade

MDPI

Special Issue Editors
Andreas Hermann
The University of Edinburgh
UK

Dominik Kurzydłowski
Cardinal Stefan Wyszyński University
Poland

Editorial Office
MDPI
St. Alban-Anlage 66
4052 Basel, Switzerland

This is a reprint of articles from the Special Issue published online in the open access journal *Crystals* (ISSN 2073-4352) in 2019 (available at: https://www.mdpi.com/journal/crystals/special_issues/ First-Principles).

For citation purposes, cite each article independently as indicated on the article page online and as indicated below:

LastName, A.A.; LastName, B.B.; LastName, C.C. Article Title. *Journal Name* **Year**, *Article Number*, Page Range.

ISBN 978-3-03921-670-3 (Pbk)
ISBN 978-3-03921-671-0 (PDF)

Contents

About the Special Issue Editors

Andreas Hermann is a Reader (associate professor) in the School of Physics and Astronomy at the University of Edinburgh, UK. He obtained his PhD in Sciences from Massey University, New Zealand, in 2010. After moving to a postdoctoral position at Cornell University, USA, in 2013 he joined the faculty at the University of Edinburgh. He works in computational materials science, with a particular focus on first-principles calculations of condensed matter systems. As member of Edinburgh's Centre for Science at Extreme Conditions, part of his research focuses on the properties of materials under extreme pressure and temperature conditions. This includes the pressure responses of molecular systems and minerals, with implications for geo- and planetary sciences, but also exploring computationally the synthesis of new materials with interesting electronic or mechanical properties, such as polyhydrides or superhard materials. He has co-authored over 70 peer-reviewed publications and two review articles on light-element superconductivity and chemical bonding under pressure.

Dominik Kurzydłowski is an associate professor at the Faculty of Mathematics and Natural Sciences at the Cardinal Stefan Wyszyński University in Warsaw. He obtained his PhD in Chemistry from the University of Warsaw in 2013. After completing his post-doctoral research stay in the group of dr. Mikhail Eremets at the Max Planck Institute of Chemistry (Mainz, Germany), in 2014 he joined the Cardinal Stefan Wyszyński University. His research interest encompass the electronic and magnetic properties of fluorides, as well as the high-pressure chemistry of these compounds. In his research he employs both experimental (Raman scattering, X-ray diffraction), and theoretical methods (solid-state density functional theory).

crystals

MDPI

Editorial

First-Principles Prediction of Structures and Properties in Crystals

Andreas Hermann [1,*] **and Dominik Kurzydłowski** [2,*]

1   Centre for Science at Extreme Conditions, School of Physics and Astronomy and SUPA, The University of Edinburgh, EH9 3FD Edinburgh, UK
2   Faculty of Mathematics and Natural Sciences, Cardinal Stefan Wyszyński University, 01-815 Warsaw, Poland
*   Correspondence: a.hermann@ed.ac.uk (A.H.); d.kurzydlowski@uksw.edu.pl (D.K.);
    Tel.: +44-131-650-5824 (A.H.); +48-22-380-96-30 (D.K.)

Received: 3 September 2019; Accepted: 3 September 2019; Published: 4 September 2019

The term "first-principles calculations" is a synonym for the numerical determination of the electronic structure of atoms, molecules, clusters, or materials from 'first principles', i.e., without any approximations to the underlying quantum-mechanical equations that govern the behavior of electrons and nuclei in these systems. In principle, these calculations allow us to learn about structural, mechanical, electronic, optical, and many more properties of molecules and materials without having to resort to any empirical or effective models. Of course, solving the quantum many-particle problem is very hard; the application of the clamped nuclei or Born–Oppenheimer approximations removes the atomic nuclei from the problem but leaves behind the many-electron problem that is still impossible to solve analytically for any interesting system.

Quantum chemists have risen to the challenge since the late 1920s by developing a succession of approximate approaches that, crucially, can be extended systematically towards solving the full many-electron problem. For small molecular systems, the best of these methods (such as coupled cluster theory) can determine electronic properties such as the ionization potential as accurately and precisely as the best experiments. These approaches cannot be scaled well to extended, crystalline systems, and it was the development of density functional theory (DFT) in the 1960s that opened the door to accurate "first-principles" calculations of crystalline materials. DFT comes with its own methodological challenges and restrictions, first and foremost that a crucial component of the electron–electron interaction, the exchange-correlation energy, is only known in a myriad of approximations and cannot be extended systematically towards the true expression. Nonetheless, DFT calculations have shown from the beginning that they provide a reasonably accurate means to reproduce and explain experimentally measured properties of crystals [1].

However, the challenge of whether first-principles calculations can evolve from an *explanatory* to a truly *predictive* tool was acknowledged, most colorfully in John Maddox's famous statement in the late 1980s about the "continuing scandal in the physical sciences"—the failure to predict the crystal structures of materials on the basis of their chemical composition alone [2]. This 'scandal' has been tackled over the last two decades: more accurate calculations, smarter algorithms that borrow from data science, biological evolution, and machine learning, all combined with increased computing power, allowing us to use DFT for truly predictive purposes [3–5].

Yet, challenges remain: complex chemical compositions, variable external conditions (such as pressure), defects, or properties that rely on collective excitations—all represent computational and/or methodological bottlenecks. This Special Issue comprises a collection of papers that use DFT to tackle some of these challenges and thus highlight what can (and cannot yet) be achieved using first-principles calculations of crystals.

In Reference [6], Geng et al. have used evolutionary algorithms to predict crystal structures and electronic properties of binary Na–S compounds under pressure. This material class is of interest for use in battery materials, but the authors show that high pressure favors the formation of metallic compounds with intriguing superconducting properties. Their work searches for stable and metastable phases across different pressures and a wide range of chemical compositions, and the authors use quantum chemical approaches and density functional perturbation theory to obtain the full picture of the new compounds' electronic structure and superconductivity.

In Reference [7], Higgins et al. present a methodological advancement of a crystal structure prediction tool based on genetic algorithms, which accounts for the magnetic as well as the crystal structure of a material by treating the localized magnetic moments as degrees of freedom on par with the atomic positions and unit cell dimensions. They demonstrate that their approach can predict complex magnetic structures that can form at the interfaces of magnets and semiconductors.

In Reference [8], Derzsi et al. predict the structural, electronic, and magnetic properties of an elusive metal halide, $AgCl_2$. They show that structure prediction can be done both in a physically biased way (following imaginary phonon modes of reasonable candidate structures) and in an unbiased way (exploring configurational space using evolutionary algorithms). The authors examine in detail the difficulties of DFT to properly capture the charge transfer and magnetic properties of this compound and place the resultant most stable structure in a wider context of metal halide compounds.

In Reference [9], Khan et al. explore the electronic and transport properties of Mn-doped ZnTe. The host material is a promising thermoelectric, and the authors show that this is also true for the doped materials, whilst doping reduces the electronic band gap towards values that are useful for optoelectronic applications. The combination of defects, magnetism, and transport properties makes this a very challenging problem.

In Reference [10], Peng et al. tackle a methodological issue of DFT: how best to describe non-local electronic correlations of the London dispersion type. They introduce a new dispersion-corrected exchange-correlation functional that relies on a single adjustable parameter, q, which is present in the exchange part of the functional. The authors use the new approach to optimize the mass densities of a test set of molecular and layered materials and show that it outperforms two standard density-based dispersion correction methods.

In Reference [11], Chen et al. study the properties of a class of metal-iridium compounds, X_3Ir, for a set of early *3d* and *4d* transition metals. They focus on physical properties such as elastic moduli and sound velocities and are able to identify trends between the different metals, relate their calculations to experimental data, and correlate them to details of the electronic structures of the compounds.

In Reference [12], Peng et al. survey a large set of possible point defects in an exemplary III–V semiconductor, InAs. These calculations necessitate the use of large supercells of the host material crystal structure, and the authors consider a wide range of possible local charge states for the different defects. How to treat such charged point defects in a fully periodic framework remains a point of interest to the community, and the present work contributes to the body of work from first-principles calculations in this area.

In summary, this Special Issue highlights several of the frontiers of first-principles calculations in crystals: the prediction of crystal structures of materials, which remains the foundation to determine any and all of their properties; treating symmetry-breaking events such as defect formation or doping, which can significantly change materials' properties; calculating collective atomic or electronic excitations, which requires perturbative approaches that relegate the standard DFT calculation to a small first step; and developing new and improved ways to capture exchange-correlation effects in many-electron systems. The present papers show that impressive progress has been made on all these frontiers to allow truly predictive first-principles studies of crystalline materials and their properties.

We would like to thank all authors who have contributed their excellent papers to this Special Issue, the large number of reviewers who provided constructive and helpful feedback on all submissions, and the editorial staff at *Crystals* for their fast and professional handling of all manuscripts during the submission process and for the help provided throughout.

Conflicts of Interest: The authors declare no conflict of interest.

References

1. Yin, M.T.; Cohen, M.L. Microscopic Theory of the Phase Transformation and Lattice Dynamics of Si. *Phys. Rev. Lett.* **1980**, *45*, 1004–1007. [CrossRef]
2. Maddox, J. Crystals from first principles. *Nature* **1988**, *335*, 201. [CrossRef]
3. Jain, A.; Ong, S.P.; Hautier, G.; Chen, W.; Richards, W.D.; Dacek, S.; Cholia, S.; Gunter, D.; Skinner, D.; Ceder, G.; et al. Commentary: The Materials Project: A materials genome approach to accelerating materials innovation. *APL Mater.* **2013**, *1*, 011002. [CrossRef]
4. Butler, K.T.; Davies, D.W.; Cartwright, H.; Isayev, O.; Walsh, A. Machine learning for molecular and materials science. *Nature* **2018**, *559*, 547–555. [CrossRef] [PubMed]
5. Oganov, A.R.; Pickard, C.J.; Zhu, Q.; Needs, R.J. Structure prediction drives materials discovery. *Nat. Rev. Mater.* **2019**, *4*, 331–348. [CrossRef]
6. Geng, N.; Bi, T.; Zarifi, N.; Yan, Y.; Zurek, E. A First-Principles Exploration of NaxSy Binary Phases at 1 atm and Under Pressure. *Crystals* **2019**, *9*, 441. [CrossRef]
7. Higgins, E.J.; Hasnip, P.J.; Probert, M.I.J. Simultaneous Prediction of the Magnetic and Crystal Structure of Materials Using a Genetic Algorithm. *Crystals* **2019**, *9*, 439. [CrossRef]
8. Derzsi, M.; Grzelak, A.; Kondratiuk, P.; Tokár, K.; Grochala, W. Quest for Compounds at the Verge of Charge Transfer Instabilities: The Case of Silver(II) Chloride. *Crystals* **2019**, *9*, 423. [CrossRef]
9. Khan, W.; Azam, S.; Ullah, I.; Rani, M.; Younus, A.; Irfan, M.; Czaja, P.; Kityk, I.V. Insight into the Optoelectronic and Thermoelectric Properties of Mn Doped ZnTe from First Principles Calculation. *Crystals* **2019**, *9*, 247. [CrossRef]
10. Peng, Q.; Wang, G.; Liu, G.-R.; De, S. Van der Waals Density Functional Theory vdW-DFq for Semihard Materials. *Crystals* **2019**, *9*, 243. [CrossRef]
11. Chen, D.; Geng, J.; Wu, Y.; Wang, M.; Xia, C. Insight into Physical and Thermodynamic Properties of X3Ir (X = Ti, V, Cr, Nb and Mo) Compounds Influenced by Refractory Elements: A First-Principles Calculation. *Crystals* **2019**, *9*, 104. [CrossRef]
12. Peng, Q.; Chen, N.; Huang, D.; Heller, E.; Cardimona, D.; Gao, F. First-Principles Assessment of the Structure and Stability of 15 Intrinsic Point Defects in Zinc-Blende Indium Arsenide. *Crystals* **2019**, *9*, 48. [CrossRef]

crystals

MDPI

Article

A First-Principles Exploration of Na_xS_y Binary Phases at 1 atm and Under Pressure

Nisha Geng [1], Tiange Bi [1], Niloofar Zarifi [1], Yan Yan [1,2] and Eva Zurek [1,*]

[1]   Department of Chemistry, University at Buffalo, Buffalo, NY 14260, USA
[2]   School of Sciences, Changchun University, Changchun 130022, China
[*]   Correspondence: ezurek@buffalo.edu

Received: 1 August 2019; Accepted: 20 August 2019; Published: 24 August 2019

Abstract: Interest in Na-S compounds stems from their use in battery materials at 1 atm, as well as the potential for superconductivity under pressure. Evolutionary structure searches coupled with Density Functional Theory calculations were employed to predict stable and low-lying metastable phases of sodium poor and sodium rich sulfides at 1 atm and within 100–200 GPa. At ambient pressures, four new stable or metastable phases with unbranched sulfur motifs were predicted: Na_2S_3 with $C2/c$ and $Imm2$ symmetry, $C2$-Na_2S_5 and $C2$-Na_2S_8. Van der Waals interactions were shown to affect the energy ordering of various polymorphs. At high pressure, several novel phases that contained a wide variety of zero-, one-, and two-dimensional sulfur motifs were predicted, and their electronic structures and bonding were analyzed. At 200 GPa, $P4/mmm$-Na_2S_8 was predicted to become superconducting below 15.5 K, which is close to results previously obtained for the β-Po phase of elemental sulfur. The structures of the most stable M_3S and M_4S, M = Na, phases differed from those previously reported for compounds with M = H, Li, K.

Keywords: high-pressure; crystal structure prediction; electronic structure; battery materials; superconductivity

---

1. Introduction

At atmospheric pressures, the size of the metal atom is thought to be important in determining the alkali metal polysulfide stoichiometries that are stable. In the case of the lightest alkali, lithium, only Li_2S and Li_2S_2 have been made [1], whereas for the heavier metal atoms (M = K, Rb, Cs) the known phases include M_2S_n with n = 1–6 [2–5]. The solid state Na-S system is particularly fascinating, with manuscripts reporting the failed synthesis of previously published stoichiometries, or the synthesis of new polymorphs. Na_2S, Na_2S_2, Na_2S_4 and Na_2S_5 are stable or metastable at atmospheric conditions [2,6–8]. Two polymorphs of Na_2S_2 are known: the α form with three formula units per cell that is stable below 170 °C, and the higher temperature β form with two formula units per cell [9]. Several studies have failed to synthesize or isolate Na_2S_3, with the reaction products yielding a mixture of Na_2S_2 and Na_2S_4 instead [6]. A novel synthesis route in liquid ammonia yielded Na_2S_3, but the product decomposed around 100 °C [10]. The crystal structure of α-Na_2S_5 has been solved [11], and several metastable polymorphs with this stoichiometry have been detected via Raman spectroscopy [2] and X-ray diffraction [8], but their structures were not determined. This amazing structural versatility is in part due to the ability of sulfur to form anionic polysulfide chains, S_n^{2-}, with various lengths. Only unbranched chains are found in the sodium polysulfides, but they may have different arrangements or conformations.

Research on the Na-S system has been motivated by Kummer and Weber's development of the battery containing these two elements at the Ford Motor Company in 1966 [12–14]. The advantages of the Na-S battery include the fact that it is made from inexpensive materials, has a long cycle lifetime, and it can deliver high-energy densities [15–18]. Density functional theory (DFT) calculations [19–21] and experiments [20] have been carried out to study the ambient pressure phases with Na_2S_n, $n > 1$. Momida and co-workers only considered the known phases. They found that the inclusion of dispersion, as implemented in Grimme's PBE-D2 functional, did not substantially influence their structural parameters nor their energies, in contrast to the results for elemental sulfur [19]. DFT calculations have concluded that, based upon the energy alone, β-Na_2S_2 is somewhat more stable than α-Na_2S_2 [19,20]. Crystal structure prediction (CSP) techniques coupled with DFT calculations have been employed to search for new ambient pressure phases [20,21]. Mali and co-workers computed the enthalpies of formation of the novel $Cmme$-Na_2S_3 and $C2$-Na_2S_6 phases, and found that they lay only slightly above the convex hull, suggesting these crystalline lattices may be metastable. A novel low energy polymorph of Na_2S_5 was predicted, and a higher energy ϵ polymorph was synthesized. Both of these were computed to be more stable than the known α phase. Wang and co-workers predicted a new polymorph, γ-Na_2S_2, which was somewhat higher in energy than the known α and β forms [21]. Moreover, a $C2/c$ symmetry Na_2S_3 structure, which contained V-shaped S_3^{2-} units, was found to be thermodynamically and dynamically stable.

The pressure induced structural transitions, as well as the electronic and optical properties of the dialkali sulfides have been studied extensively. At ambient conditions, Li_2S, Na_2S, K_2S and Rb_2S assume the antifluorite (anti-CaF_2) structure [22,23], whereas Cs_2S crystallizes in an anticottunite (anti-$PbCl_2$) structure [24]. Li_2S and Na_2S transform to an isotypic anticottunite phase at 12 GPa [25], and 7 GPa [26]. A further transition to an Ni_2In phase occurs at 30–45 GPa [27] and 16 GPa [26] in these systems, respectively. The sequence of phase transitions for Rb_2S are: anti-$CaF_2 \rightarrow$ anti-$PbCl_2 \rightarrow Ni_2In$ at 0.7 and 2.6 GPa [28]. Cs_2S assumes a distorted Ni_2In structure by 5 GPa [29]. Many of these phase transitions were either predicted or confirmed via theoretical calculations [30]. Recently, CSP has been used to search for the most stable structures of Li_2S, Na_2S [31], and K_2S at higher pressures [31,32]. Na_2S was predicted to transform into an anti-AlB_2 phase at 162 GPa, and an anti-KHg_2 structure at 232–244 GPa [31].

The discovery of conventional superconductivity in a hydride of sulfur that has been identified as $Im\bar{3}m$-H_3S has invigorated the quest for high-temperature superconductors with unique stoichiometries [33]. The superconducting critical temperature, T_c, measured for this material was a record breaking 203 K at 150 GPa [34]. Because the valence shells of the alkali metal hydrides are isoelectronic with hydrogen, Kokail and co-workers hypothesized that the alkali sulfides could also be good superconductors [35]. CSP was employed to search for stable structures in the metal rich Li-S [35], and K-S [36] phase diagrams under pressure, however most of the phases found had no or low T_c. The only exception was a $Fm\bar{3}m$-Li_3S structure whose T_c reached a maximum of 80 K at 500 GPa, a pressure at which it was metastable [35].

Herein, we carry out a comprehensive theoretical investigation that employs an evolutionary structure search to predict the most stable, and low-lying metastable phases in the metal rich and metal poor regions of the Na-S phase diagram at 1 atm, as well as 100–200 GPa. In addition to identifying many of the known or previously predicted ambient pressure phases, novel polymorphs with the Na_2S_3, Na_2S_5, and Na_2S_8 stoichiometries, which could potentially be synthesized, are found, and their properties are reported. Dynamically stable Na_3S and Na_4S phases, whose structures are related to the antifluorite Na_2S phase, lie ~70 meV/atom above the 0 GPa convex hull. (Meta)stable structures in the sulfur rich region of the phase diagram at 100 GPa contain a wide range of sulfur motifs including: zero-dimensional dimers or trimers, one-dimensional zigzag or branched tertiary chains, as well as fused square or cyclohexane motifs. By 200 GPa, most of the predicted phases are comprised of two-dimensional square nets or cubic-like lattices. The most stable Na_3S and Na_4S phases at 200 GPa did not bear any resemblance to H_3S [37], or the sulfides of lithium [35] and potassium [36] whose superconducting properties have previously been investigated. We hope the structural diversity of the phases predicted herein inspires the directed synthesis of new sulfides of sodium.

2. Computational Details

Evolutionary structure searches were carried out to find stable and low-lying metastable crystals in the Na-S phase diagram: in the sulfur rich region the Na_2S_n stoichiometry with n = 2–6, 8, and in the metal rich region the Na_nS stoichiometry with n = 2–4 were considered. The calculations were carried out using the open-source evolutionary algorithm (EA) XTALOPT [38,39] version 10 [40], wherein duplicate structures were removed from the breeding pool via the XTALCOMP algorithm [41], and random symmetric structures were created in the first generation using RANDSPG [42]. EA searches were performed on structures containing 1–4 formula units in the primitive cell at 0, 100, 150 and 200 GPa. Minimum interatomic distance constraints were chosen to generate the starting structures in each generation. The minimum distances between Na-Na, Na-S and S-S atoms were set to 1.86, 1.45, and 1.04 Å, respectively. To improve the efficiency of the CSP searches, and increase the size of the unit cell that could be considered, the evolutionary search was seeded with experimentally determined and theoretically predicted structures from the literature, when available. The most stable structures found in the high pressure EA searches were optimized between 100 and 200 GPa, and their relative enthalpies and equations of states (EOS) are provided in the Supplementary Materials.

Geometry optimizations and electronic structure calculations including the densities of states (DOS), band structures, electron localization functions (ELFs) and Bader charges were carried out using DFT as implemented in the Vienna *Ab-Initio* Simulation Package (VASP) [43,44]. The bonding of select phases was further analyzed by calculating the crystal orbital Hamilton populations (COHP) and the negative of the COHP integrated to the Fermi level (-iCOHP) using the LOBSTER package [45]. At all pressures, the gradient-corrected exchange and correlation functional of Perdew–Burke–Ernzerhof (PBE) [46] was employed. It has been shown that it is important to include van der Waals (vdW) interactions to obtain reasonable estimates for the volume of α-S at 1 atm [19]. Therefore, the most stable structures from the 0 GPa PBE EA searches were reoptimized with the optB88-vdW functional [47,48]. In the EA searches, we employed plane-wave basis set cutoff of 325–400 eV, and the k-point meshes were produced by the Γ-centered Monkhorst–Pack scheme with the number of divisions along each reciprocal lattice vector chosen so that the product of this number with the real lattice constant was 30 Å. This value was increased to 50 Å for precise optimizations. The atomic potentials were described using the projector augmented wave (PAW) method [49]. The S $3s^23p^4$ electrons were treated explicitly in all of the calculations. At 0 GPa the Na $3s^1$ valence electron configuration was used, and at higher pressures an Na $2s^22p^63s^1$ valence configuration was employed. For precise optimizations, the plane-wave basis set cutoffs were increased to 700 eV at 0 GPa, and 1000 eV at high pressures.

To verify the dynamic stability phonon band structures were calculated via the supercell approach [50,51], wherein the dynamical matrices were calculated using the PHONOPY code [52]. The electron–phonon coupling (EPC) calculations were performed using the Quantum Espresso (QE) program [53]. The Na $(2s^22p^63s^1)$ and S $(3s^23p^4)$ pseudopotentials, obtained from the PSlibrary [54], were generated by the method of Trouiller–Martins [55] with the PBE functional, and an energy cutoff of 90 Ry was chosen. The Brillouin zone sampling scheme of Methfessel–Paxton [56] and 24 × 24 × 6 k-point grid and a 8 × 8 × 2 q-point grid were used for $P4/mmm$ Na_2S_8 at 200 GPa. The EPC parameter, λ, was calculated using a set of Gaussian broadenings with an increment of 0.02 Ry from 0.0 to 0.600 Ry. The broadening for which λ was converged to within 0.05 Ry was 0.10 Ry. T_c was estimated using the Allen–Dynes modified McMillan equation [57] with a renormalized Coulomb potential, μ^*, of 0.1.

3. Results and Discussion

3.1. Stable and Metastable Na-S Phases at Atmospheric Conditions

The enthalpies of formation, ΔH_F, of the most stable Na-S phases found in our EA searches are plotted in Figure 1 as a function of pressure. The phases whose ΔH_F lie on the convex hull are thermodynamically stable, while those whose ΔH_F are not too far from the hull may be metastable

stable provided their phonon modes are all real. At atmospheric pressures, calculations carried out with both the PBE and optB88-vdW functionals showed that the Na_2S, Na_2S_2, Na_2S_4 and Na_2S_5 stoichiometries lay on the hull. The ΔH_F of Na_2S_3, Na_2S_6 and Na_2S_8 were slightly above the hull, whereas those of Na_3S and Na_4S were further away from it. The inclusion of vdW interactions lowered the total ΔH_F by no more than 80 meV/atom, but the stoichiometries that lay on the hull were the same as those found within PBE. The inclusion of the zero point energy, ZPE, increased the ΔH_F by no more than 9 meV/atom, but it also did not affect the identities of the thermodynamically stable phases. The structural parameters and ΔH_F of the stable and important metastable phases are provided in the Supplementary Materials. All of the structures identified in the EA search that were within 15 meV/atom of the lowest enthalpy geometry for a particular stoichiometry were examined, and none of them contained sulfur anions with branched chains.

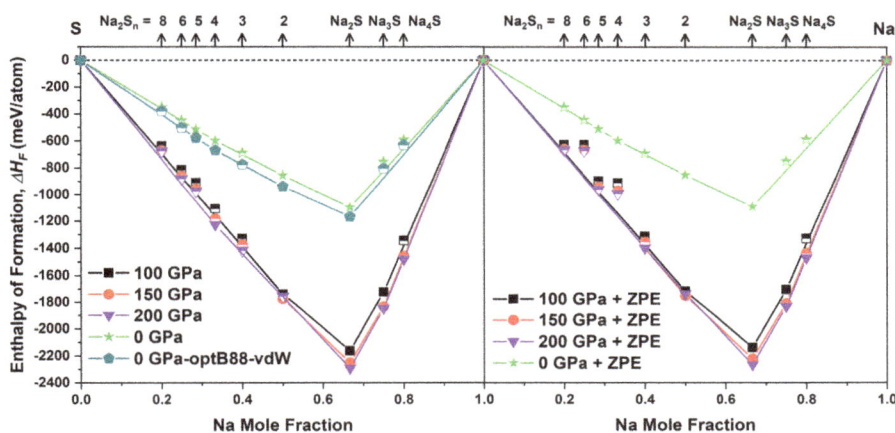

Figure 1. (**left**) Enthalpy of formation, ΔH_F, of the Na_xS_y compounds with respect to solid Na and S as computed with the optB88-vdW functional at 0 GPa and the PBE functional between 0–200 GPa. (**right**) PBE results including the zero point energy (ZPE). Closed symbols lie on the convex hull (denoted by solid lines), open symbols lie above it. ΔH_F was calculated using the enthalpies of the experimentally known structures: body-centered cubic (bcc, 0 GPa), face-centered cubic (fcc, 100 GPa) [58], $cI16$ (150 GPa) [59,60], and $hp4$ (200 GPa) [61] for Na, and α-S (0 GPa) [62]. Because the crystal structure of S above 83 GPa is still disputed [63,64], the β-Po phase [65] was employed between 100 and 200 GPa, as in recent studies [35].

Momida [19] and Wang et al. [21] studied the effect of vdW interactions on the unit cell volumes and formation enthalpies of the sodium polysulfides. The effects of dispersion were approximated by using Grimme's semi-empirical DFT-D2 method [66] in conjunction with the PBE functional. Both studies found that vdW interactions resulted in slightly smaller volumes for the Na-S compounds, and more negative formation enthalpies. However, the inclusion of dispersion was crucial so that reasonable estimates of the unit cell volume of elemental α-S, which is composed of molecular S_8 rings, could be calculated [19]. Herein, the optB88-vdW method [47,48], which employs a non-local correlation functional that approximately accounts for dispersion interactions, was used. It has been demonstrated that this functional is among those that provides the best agreement with experiment for the volumes and lattice constants of layered electroactive materials for Li-ion batteries [67], as well as a broad range of metallic, covalent and ionic solids [48]. Other choices that might provide even more accurate lattice parameters at ambient pressures include combining PBE [46], and PBEsol [68] or its improvements [69] with rVV10 [70,71], or SCAN+rVV10 [72]. A comparison of the calculated volumes per atom obtained via PBE and optB88-vdW are provided in the Supplementary Materials. For elemental sulfur, PBE yields a cell volume of 37.21 Å^3/atom, which is 36.4% larger than the experimental volume of 25.76 Å^3/atom [73]. The optB88-vdW functional yields a volume of 24.90 Å^3/atom (c.f. 26.88 Å^3/atom with PBE-D2 [19]), which is only 3.4% lower than experiment. In Na-S systems that contained at least 50 mole % sodium, the difference between the PBE and optB88-vdW volumes was <4%, otherwise the difference ranged from 5–16%, depending on the stoichiometry and polymorph. For Na$_2$S, α-Na$_2$S$_2$, β-Na$_2$S$_2$, Na$_2$S$_4$, and Na$_2$S$_5$ optB88-vdW yielded volumes of 22.77, 21.51, 21.90, 21.85, and 22.38 Å^3/atom, respectively, which differs by <4% from Momida's PBE-D2 results [19]. The 0 K optB88-vdW ΔH_F values for the most stable Na$_2$S, Na$_2$S$_2$, Na$_2$S$_4$, and Na$_2$S$_5$ polymorphs were calculated to be -1.17, -0.94, -0.67, and -0.58 eV/atom, which is in good agreement with the experimental ΔH_F^0 values at 298.15 K of -1.26, -1.03, -0.71, and -0.61 eV/atom, respectively [74].

The anti-CaF$_2$ Na$_2$S structure with $Fm\bar{3}m$ symmetry was the lowest point on the 0 GPa convex hull [19–21], and the Na$_2$S$_2$ stoichiometry had the second most negative ΔH_F. Our EA searches were seeded only with the known α and β-Na$_2$S$_2$ polymorphs, but they also readily identified a higher energy γ-Na$_2$S$_2$ phase that has recently been predicted [21]. All of these three polymorphs are comprised of Na$^+$ cations and S$_2^-$ anions. In agreement with previous DFT calculations [19–21], the β polymorph was computed to have the lowest ΔH_F, followed by the α, and the γ configurations. The PBE/optB88-vdW differences in energy between the α and β structures were comparable to the difference between the β and γ structures, 4/8 meV/atom and 5/9 meV/atom, respectively.

In addition to Na$_2$S and Na$_2$S$_2$, the Na$_2$S$_4$ and Na$_2$S$_5$ stoichiometries also lay on the PBE and optB88-vdW convex hulls. The EA search was seeded with the known $I\bar{4}2d$-Na$_2$S$_4$ structure [75], which contains an unbranched S$_4^{2-}$ chain whose S-S bond angle and dihedral angle were computed to be 111.3° and 96.7°, within PBE, respectively. No other polymorphs with comparable energies were found.

The EA search was also seeded with the α-Na$_2$S$_5$ [11] and ϵ-Na$_2$S$_5$ [20] polymorphs illustrated in Figure 2a,b. In the α form, the unbranched S$_5^{2-}$ anion adopts a bent (*cis*) configuration, whereas, in the ϵ form, it is stretched (*trans*), as in K$_2$S$_5$ [76], Rb$_2$S$_5$ [77] and Cs$_2$S$_5$ [78]. PBE and PBE-D2 calculations have shown that neither the α nor the ϵ phases lay on the convex hull [19,20], but CSP has found another currently unsynthesized phase with stretched S$_5^{2-}$ anions that was thermodynamically stable [20]. The coordinates of this phase were not provided in Réf. [20], however it appears to be different from the lowest enthalpy C2 symmetry Na$_2$S$_5$ phase from our EA searches shown in Figure 2c. In the structure of Mali and co-workers [20], neighboring S$_5^{2-}$ chains point in opposite directions (similar to what is observed in the ϵ phase along the b-lattice vector), whereas in C2-Na$_2$S$_5$ they face the same direction. C2-Na$_2$S$_5$ lay on the PBE convex hull, and it's enthalpy was 4 and 5 meV/atom lower than the ϵ and α polymorphs, respectively. The order of stability was not affected by the ZPE contributions. On the other hand, within optB88-vdW, the ϵ phase, which lay on the convex hull, had the lowest enthalpy with the α and C2 phases being 3 and 25 meV/atom higher, respectively. These results suggest that other energetically competitive polymorphs, based upon unbranched S$_5^{2-}$ units with either *cis* or *trans* geometries, could potentially be constructed, and that the computed energy ordering depends upon the method used to treat dispersion. PBE calculations showed that all three Na$_2$S$_5$ polymorphs had indirect band gaps with values of 1.73, 1.47, and 1.30 eV for the ϵ, α, and C2 phases, respectively (see the Supplementary Materials). Better estimates could be obtained using hybrid density functionals or GW, however the PBE results suggest that the conformation of the S$_5^{2-}$ anion and the geometry of the cell both have an effect on the band gap.

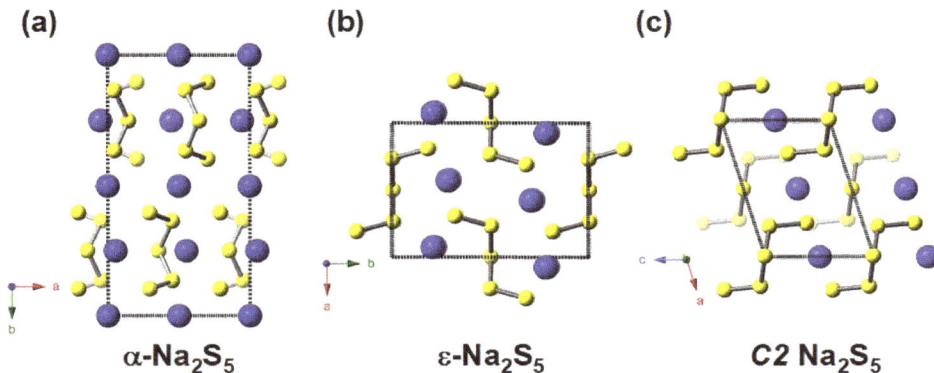

Figure 2. Unit cells of the previously known (**a**) α-Na$_2$S$_5$ [11] and (**b**) ϵ-Na$_2$S$_5$ [20] polymorphs, as well as the newly predicted (**c**) C2-Na$_2$S$_5$ structure. Sodium atoms are colored blue, and sulfur atoms are yellow.

Besides the previously mentioned thermodynamically stable stoichiometries, several sulfur-rich containing phases, Na$_2$S$_n$ ($n = 3, 6, 8$), were found to be low-energy metastable species, as confirmed by phonon calculations. For the Na$_2$S$_3$ stoichiometry, the *Cmme*, *C2/c*-I, *C2/c*-II, and *Imm*2 polymorphs illustrated in Figure 3 lay 7/22, 3/1, 13/9, and 17/40 meV/atom above the convex hull within PBE/optB88-vdW calculations. CSP previously predicted the *Cmme* [20] and *C2/c*-I [21] structures, used as seeds in our EA searches, whereas the *C2/c*-II and *Imm*2 phases discovered here have not been reported before. All four of these polymorphs contained V-shaped S$_3^{2-}$ anions with S-S bond lengths of 2.087–2.107 Å and bond angles of 106.00–111.21°. The main difference between them was the relative arrangement of the S$_3^{2-}$ motifs. In *Imm*2 all of the V-shaped anions pointed in the same direction, whereas in *C2/c*-II those in a single layer pointed in the same direction, but those in the adjacent layer were rotated by 180°. In *Cmme* and *C2/c*-I adjacent rows of anions in the same layer pointed in opposite directions. In *Cmme* the apex of the Vs in one layer were located directly behind those in an adjacent layer, but rotated by 180°. In *C2/c*-I, the Vs in adjacent layers also faced opposite directions. All four polymorphs had indirect band gaps, with the PBE value for *C2/c*-I, 1.06 eV, being about 0.5 eV smaller than for the other three. The closeness of the ΔH_F of these four polymorphs to the convex hull, and their dynamic stability suggests that they may be synthesizable. Experimentalists have not yet been able to synthesize a persistent Na$_2$S$_3$ compound, yielding a mixture of Na$_2$S$_4$ and Na$_2$S$_5$ either directly [6,20,79] or after disproportionation near 100 °C [10], suggesting that the kinetic barriers towards decomposition may be low.

Seed structures were not employed in EA searches carried out on the Na$_2$S$_6$ stoichiometry, and the two nearly isoenthalpic C2 and $P\bar{1}$ symmetry structures illustrated in Figure 4a,b were found. Whereas the former was comprised of unbranched S$_6^{2-}$ units, the latter contained V-shaped trimers. These two dynamically stable phases lay only 2/3 and 5/16 meV/atom above the PBE/optB88-vdW convex hulls, respectively. The inclusion of the ZPE decreased the PBE difference in energy to 1 meV/atom. Mali and co-workers found the same two structures in their CSP searches [20]. The calculated phonon spectrum of $P\bar{1}$-Na$_2$S$_6$ contained two bands above 500 cm^{-1} (565 and 543 cm^{-1} at Γ), which is higher than in any of the other 0 GPa phase considered herein. The S-S bond distances in $P\bar{1}$-Na$_2$S$_6$ were shorter than in any of the predicted Na$_2$S$_3$ phases, 1.972 and 2.048 Å, giving rise to the increased frequency. The decreased distance is likely a result of the smaller formal charge on the trimer, S$_3^-$ in $P\bar{1}$-Na$_2$S$_6$ vs. S$_3^{2-}$ in the Na$_2$S$_3$ polymorphs in Figure 3. Molecular calculations using the ADF program package [80,81] with a triple-zeta basis set with polarization functions (TZP), and the PBE functional yielded S-S bond lengths of 2.037 and 2.166 Å in the monoanion and dianion, respectively, which is in reasonable agreement with the periodic calculations. Both phases were computed to have indirect PBE band gaps, 1.75 eV in the *C*2 and 0.72 eV in the *P*$\bar{1}$ phases, respectively.

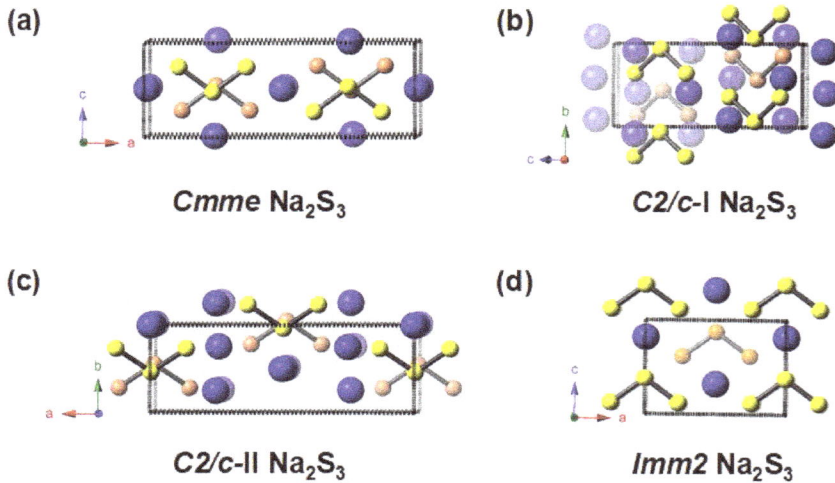

Figure 3. Unit cells of the previously predicted (**a**) *Cmme*-Na$_2$S$_3$ [20], and (**b**) *C*2/*c*-I Na$_2$S$_3$ [21] phases, as well as the newly found (**c**) *C*2/*c*-II Na$_2$S$_3$, and (**d**) *Imm*2-Na$_2$S$_3$ geometries. Sodium atoms are colored blue, whereas sulfur atoms in adjacent layers are yellow and orange.

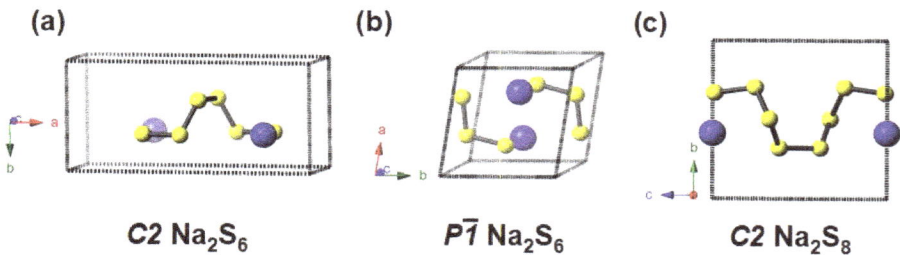

Figure 4. Unit cells of the previously predicted (**a**) *C*2-Na$_2$S$_6$ [20], and (**b**) *P*$\bar{1}$-Na$_2$S$_6$ polymorphs. (**c**) Unit cell of the newly predicted *C*2-Na$_2$S$_8$ phase. Sodium atoms are colored blue, and sulfur atoms are yellow.

Finally, EA searches were carried out on the Na$_2$S$_8$ stoichiometry, which has not been considered in other studies before. The dynamically stable *C*2 symmetry structure shown in Figure 4c was found to have the lowest energy and it lay 10/30 meV/atom above the PBE/optB88-vdW convex hulls. This phase is comprised of an unbranched S$_8^{2-}$ stretched, helical-like chain. Its PBE band gap was indirect and measured 1.56 eV.

Two sodium rich metastable phases, whose structures can be derived from the known 1 atm *Fm*$\bar{3}$*m*-Na$_2$S configuration, which has one formula unit per primitive cell and is shown in Figure 5a, were found. The *P*$\bar{3}$*m*1-Na$_3$S and *Pmn*2$_1$-Na$_4$S phases illustrated in Figure 5b,c lay 69/71 and 67/70 meV/atom above the PBE/optB88-vdW convex hull, respectively. In the antifluorite Na$_2$S structure the S^{2-} anions are arranged on a face centered cubic (fcc) lattice, and all of the tetrahedral sites are filled with Na$^+$ cations, with the Na-Na nearest neighbor distances measuring 3.267 Å. The *P*$\bar{3}$*m*1-Na$_3$S phase can be derived from a three-formula unit supercell of the antifluorite structure where one sulfur atom has been removed from the fcc lattice. This results in a distortion from perfect cubic symmetry, with Na-Na-Na angles measuring 89.3 and 89.9°, and Na-Na nearest neighbor distances measuring 3.288 and 3.242 Å. Another dynamically stable $R\bar{3}m$ symmetry phase that was 0.7/1.5 meV/atom less stable within PBE/optB88-vdW was also found in the structure searches. Its coordinates and computed properties are provided in the Supplementary Materials. The $Pmn2_1$-Na_4S structure contained four formula units. It can be obtained by inserting layers of sodium atoms into a supercell of the antifluorite structure, which results in a minimal structural distortion in the anti-CaF$_2$ layers, with Na-Na distances along the a and c lattice vectors measuring 3.344 and 3.267 Å, respectively.

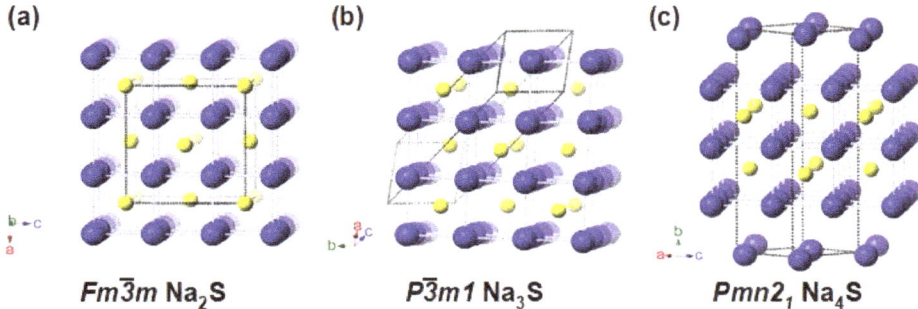

Figure 5. Crystal structures of (**a**) $Fm\bar{3}m$-Na_2S, (**b**) $P\bar{3}m1$-Na_3S and (**c**) $Pmn2_1$-Na_4S phases at 0 GPa. Sodium atoms are colored blue, and sulfur atoms are yellow.

The PBE band gap in $Fm\bar{3}m$-Na_2S was computed to be 2.47 eV, as expected for an ionic phase. As shown in the Supplementary Materials, the sodium rich $P\bar{3}m1$-Na_3S and $Pmn2_1$-Na_4S compounds are metals. Plots of the ELF, given in Figure 6, illustrate that regions where the bonding is metallic, indicated by an ELF value of ∼0.5, are localized along the two-dimensional layers of sodium atoms. The DOS of both structures between −2.5 eV to the Fermi level has step-like features, as would be expected for a two-dimensional electron gas [82]. The ELF plots also show spherical regions with high ELF values that encompass the sulfur atoms, suggestive of an S^{2-} oxidation state, and regions with a high ELF value between adjacent Na layers within $Pmn2_1$-Na_4S . The latter are labeled "Es" in Figure 6d, since they resemble interstitial electrons that are paired in anion-like species, which have been computed for a number of high-pressure electrides exhibiting ionic bonding [83,84].

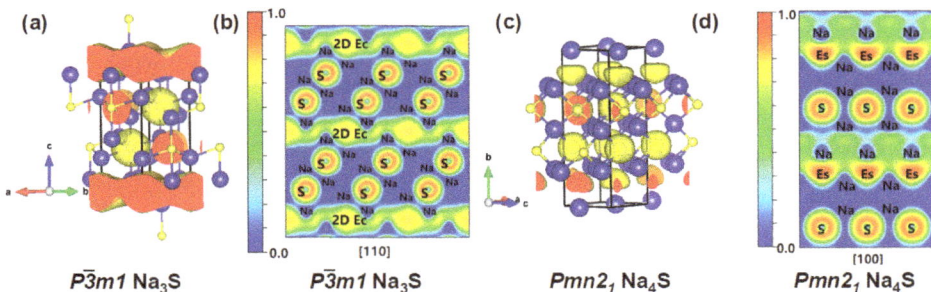

Figure 6. The electron localization function (ELF) plots of sodium rich phases at 0 GPa. $P\bar{3}m1$-Na_3S: (**a**) ELF isosurface plot using an isovalue of 0.5, and (**b**) slice of the ELF in the [100] plane. $Pmn2_1$-Na_4S: (**c**) ELF isosurface plot with an isovalue of 0.8, and (**d**) slice of the ELF in the [110] plane. In the two-dimensional plots, the positions of the S and Na atoms are denoted, as is the plane of the two-dimensional electron cloud (Ec) and the positions of the interstitial electron pairs (Es).

3.2. Na-S System at High Pressure

Because dispersion interactions are unlikely to be important under pressure [85], calculations at 100, 150 and 200 GPa were carried out with the PBE functional. Figure 1 illustrates that only Na_3S and Na_2S were thermodynamically stable at all of these pressures. Na_2S_2 lay on the 100 and 150 GPa convex hulls, whereas Na_4S and Na_2S_4 lay on the 200 GPa convex hull. The inclusion of the ZPE corrections did not affect the identity of the thermodynamically stable stoichiometries at 100 and 150 GPa, but it added Na_2S_2 and Na_2S_3 to the 200 GPa hull, and pushed Na_2S_4 away from it. Because phonon calculations (see the Supplementary Materials) confirmed that all of these phases were dynamically stable, and high-pressure experiments can lead to the formation of metastable species [86]; the structures of all of these phases are discussed below.

The Na_2S stoichiometry was the lowest point on the convex hull among all other binary phases between 100–200 GPa. The species seeded into the EA, including the experimentally known Ni_2In structure ($P6_3/mmc$) [26] and a previously predicted anti-AlB_2 structure ($P6/mmm$) [31] were found to be the most stable at 100–150 GPa and 200 GPa, respectively. DFT calculations have predicted that Na_2S is an insulator up to 300 GPa, with PBE band gaps of ∼3.50, 3.25, and 1.70 eV at 100, 150 and 200 GPa, respectively [31].

The evolutionary search for Na_2S_2 at 100–200 GPa found a *Pbam* symmetry structure, shown in Figure 7a, which contained two formula units in its primitive cell. It lay on the convex hull at 100 and 150 GPa, and was only 5 meV/atom away from the hull at 200 GPa. Its shortest S-S distances were no longer than 2.040 Å at all of the pressures studied, which is comparable to the measured S-S bond length within α-S at ambient conditions, 2.055 Å [73]. A plot of the ELF, shown in the Supplementary Materials, confirmed the presence of layers of S_2^{2-} dimers, arranged in a herringbone type fashion.

The Na_2S_3 stoichiometry lay above the convex hull at all pressures considered, however the distance from the hull decreased from 65 to 25 meV/atom from 100 to 200 GPa. The $C2/c$ phase shown in Figure 7b, which was metastable between 100 and 150 GPa, is comprised of one-dimensional branched tertiary sulfur chains with S-S distances of 2.180 Å along the chain, and 2.030 Å along the branch. By 150 GPa, the $I4/mmm$ structure illustrated in Figure 8a, was enthalpically preferred. It is comprised of two layers of square sulfur nets, with S-S distances of 2.092 Å at 200 GPa, separated by sodium nets.

Na_2S_4 was thermodynamically stable at 200 GPa, but it lay 55 and 3 meV/atom from the 100 and 150 GPa hulls. The most stable structure between 100–110 GPa, shown in Figure 7c, had $C2/c$ symmetry and contained layers of hexagonal puckered corner-sharing chair cyclohexane-like rings with S-S distances of 2.105 and 2.232 Å. The Bader charges on the 4- and 2-coordinated sulfur atoms were $-0.16e$ and $-0.60e$, respectively. Above 110 GPa, the $I4/mmm$ structure, shown in Figure 8b, was preferred. Similar to $I4/mmm$-Na_2S_3, it was comprised of two sets of two-dimensional square sulfur layers with S-S distances of 2.092 Å at 200 GPa separated by a single square net of sodium atoms.

The stoichiometries with an even larger ratio of sulfur to sodium lay between 35–85 meV/atom from the non-ZPE corrected convex hull. The structure dubbed $C2/m$-I Na_2S_5 was the most stable at 100 GPa, but by 130 GPa another phase with the same symmetry, $C2/m$-II Na_2S_5, became preferred (see Figure 7d,e). At 100 GPa, the sulfur sublattice in the former contained zigzag chains with S-S distances of 2.076 Å, as well as one-dimensional corner sharing squares with S-S distances of 2.194 Å running along the b-lattice vector. The higher pressure structure was comprised of these same one-dimensional square nets fused at the edges with S-S distances ranging from 2.029 to 2.070 Å at 200 GPa.

At 100 GPa, the lowest enthalpy Na_2S_6 structure adopted C2 symmetry (see Figure 7f), and it lay 60 meV/atom above the convex hull. It was comprised of sulfur atoms as well as dimers and V-shaped trimers, as shown in the ELF plots in the Supplementary Materials, with S-S bond lengths of 2.062 Å, and 1.998–2.180 Å, respectively. The $I4/mmm$ phase illustrated in Figure 8c lay 35 and 45 meV/atom above the hulls at 150 and 200 GPa, respectively. It resembled the high-pressure Na_2S_4

and Na_2S_3 phases, except that its sulfur layers contained both edge-sharing cubes and square nets with S-S distances of 1.916–2.111 Å and 2.111 Å, respectively at 200 GPa.

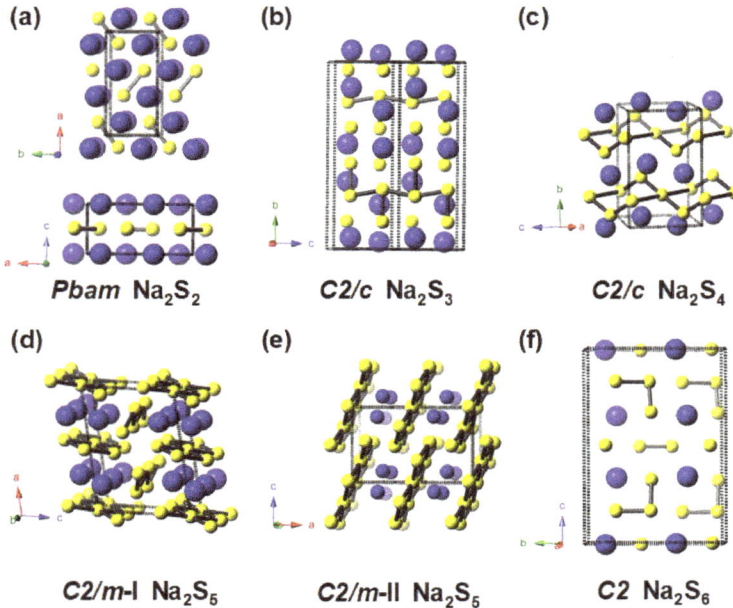

Figure 7. Predicted crystal structures of Na_2S_n ($n = 2$–6) phases under pressure whose sulfur sublattices were comprised of molecular (dimers, trimers) or one-dimensional (chains, corner or edge sharing squares, or corner sharing cyclohexane-like rings) motifs: (**a**) top and side view of *Pbam*-Na_2S_2 at 200 GPa; (**b**) $C2/c$-Na_2S_3 at 100 GPa; (**c**) $C2/c$-Na_2S_4 at 100 GPa; (**d**) $C2/m$-I Na_2S_5 at 100 GPa; (**e**) $C2/m$-II Na_2S_5 at 200 GPa; and (**f**) $C2$-Na_2S_6 at 100 GPa. Sodium atoms are colored blue, and sulfur atoms are yellow.

Evolutionary searches carried out at 100, 150, and 200 GPa for Na_2S_8 discovered the metastable $P4/mmm$ phase shown in Figure 8d. This species was reminiscent of $I4/mmm$-Na_2S_6 with layers comprised of fused cubes and square nets with S-S distances of 1.903–2.124 Å and 2.124 Å, respectively, at 200 GPa. ELF plots for $P4/mmm$-Na_2S_8 are provided in the Supplementary Materials because they are representative of the ELFs obtained for phases that contained square and/or cube-like sulfur nets. In all of the structures studied, the ELF plots suggested that the S-S bonds within the two-dimensional sheets were weaker than the bonds between atoms in different sheets. For example, in $P4/mmm$-Na_2S_8 the S-S distances along the *a*- and *b*-directions were 2.124 Å and the negative of the crystal orbital Hamilton populations integrated to the Fermi level (-iCOHPs) were 2.5 eV. For the S-S bonds that were oriented along the *c*-direction these values became 1.903 Å and 5.5 eV, which is comparable with a bond length of 2.168 Å and -iCOHP of 4.8 eV in the S_2^{2-} dimer in α-Na_2S_2 at 0 GPa. The pressure induced polymerization and bond weakening observed in the two-dimensional sulfur layers has been seen in other systems before. For example, it has been shown that the Cl_2 molecules present in XeCl at 40 GPa undergo polymerization to form one-dimensional zigzag chains by 60 GPa with the -iCOHPs between nearest neighbor atoms computed to be 3.81 eV and 2.16 eV, respectively [85]. By 100 GPa, however, there is no evidence of Cl-Cl bond formation with computed ELF and -iCOHPs characteristic for monoatomic Cl.

Within the 100–200 GPa pressure range, the most stable Na_3S stoichiometry unearthed contained two formula units in its primitive cell, and it possessed $P6_3/mmc$ symmetry. This structure, illustrated in Figure 9a, consisted of an ABAB... stacked triangular net where each sulfur atom was [6+6] coordinated with Na-S distances of 2.216 and 2.286 Å at 200 GPa. We also optimized Na_3S stoichiometries that were analogous to the $R\bar{3}m$ and $Im\bar{3}m$ structures proposed for superconducting H_3S [37]; as shown in the Supplementary Materials, their enthalpies were at least 600 meV/atom and 1000 meV/atom higher, respectively, between 100–200 GPa. Moreover, sodium analogs of the previously predicted $Pm\bar{3}m$, $I4/mmm$ and $Fm\bar{3}m$-Li_3S [35] and $Pm\bar{3}m$, $I4/mmm$-K_3S [36] phases were relaxed, and they were found to be at least 160 meV/atom less stable than $P6_3/mmc$-Na_3S between 100 and 200 GPa.

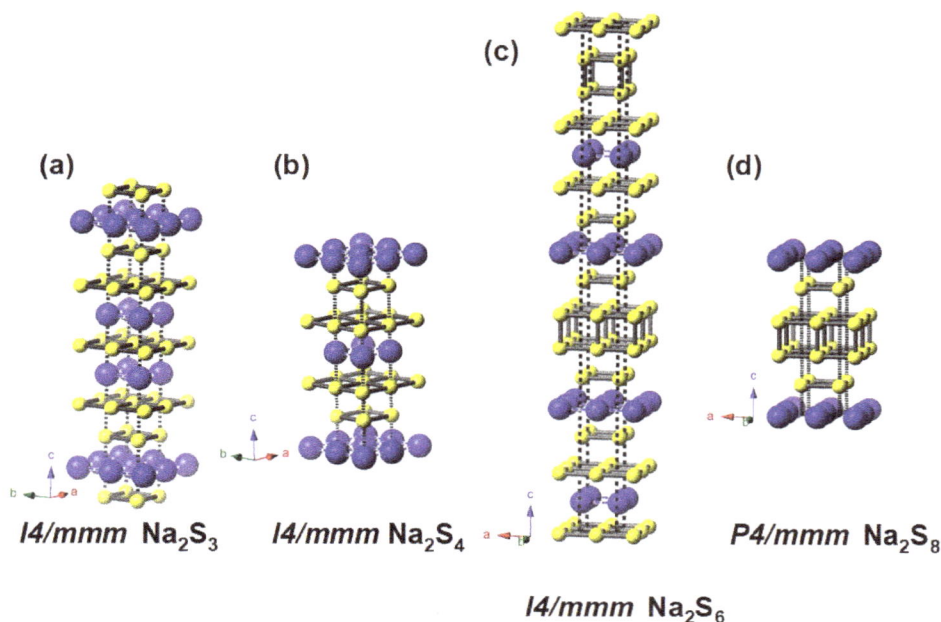

Figure 8. Predicted crystal structures of Na_2S_n ($n = 3, 4, 6, 8$) phases at 200 GPa whose sulfur sublattices were comprised of two-dimensional square or cube nets, and whose sodium sublattices were comprised of square nets: (**a**) $I4/mmm$-Na_2S_3; (**b**) $I4/mmm$-Na_2S_4; (**c**) $I4/mmm$-Na_2S_6; and (**d**) $P4/mmm$-Na_2S_8. Sodium atoms are colored blue, and sulfur atoms are yellow.

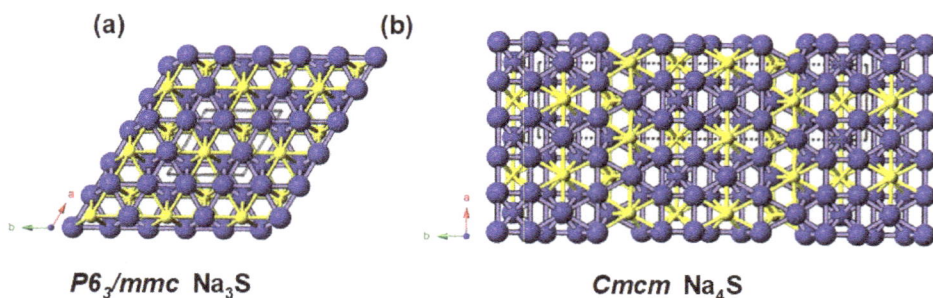

Figure 9. Predicted crystal structures of sodium rich phases at 200 GPa: (**a**) $P6_3/mmc$-Na_3S; and (**b**) $Cmcm$-Na_4S. Sodium atoms are colored blue, and sulfur atoms are yellow. Blue and yellow lines represent nearest neighbor contacts.

The *Cmcm* symmetry Na$_4$S phase, shown in Figure 9b, was also found to be the most stable in the whole pressure range of 100–200 GPa. At 200 GPa, it lay on the convex hull, but at 100 and 150 GPa, it was 35 and 10 meV/atom above the hull. It can be viewed as a cut triangular two-dimensional lattice that is joined to another such lattice via a square net. Individual sheets are stacked in an ABAB fashion. Geometry optimizations of the sodium analog of *Cmcm*-K$_4$S [36] were between 25 and 50 meV/atom less stable than *Cmcm*-Na$_4$S, whereas the enthalpies of the sodium analogs of *R$\bar{3}$m*-Li$_4$S [35] and K$_4$S [36] were at least 200 meV/atom higher than the most stable structure found here.

3.3. Electronic Structure and Superconductivity under Pressure

The PBE band structures and electronic DOS plots of the high-pressure dynamically stable phases, provided in the Supplementary Materials, showed that within their range of stability they were all metallic. Under pressure elemental sodium transforms from the ambient pressure bcc structure to an fcc phase at 65 GPa [58], followed by the semi-metallic *c*I16 structure at 103 GPa [87]. Near the minimum of the melting temperature a number of complex crystal structures, with unit cells as large as 512 atoms, have been identified experimentally within a narrow temperature/pressure range [59]. Remarkably, as first predicted by theory [88], sodium undergoes a metal to insulator (MIT) transition by 200 GPa, which results from the overlap of core electrons and the concomitant localization of $p - d$ hybridized valence electrons, which render this phase an electride [61]. Theoretical calculations have found that the maximum electron–phonon-coupling parameter, λ, for sodium occurs for the *c*I16 phase near 140 GPa [89]. Estimates of the T_c within the Allen–Dynes modified McMillan equation and $\mu^* = 0.13$ found a maximum T_c of 1.2 K, leading to the conclusion that superconductivity in sodium prior to the MIT is weak or non-existent. In addition, because DFT calculations showed that most of the metal rich binary lithium and potassium sulfides had no or low T_c [35,36], we thought it unlikely that the *P*6$_3$/*mmc*-Na$_3$S and *Cmcm*-Na$_4$S phases found herein would be good superconductors.

In contrast, compressed sulfur is among the elemental phases with the highest T_c, reaching up to 17 K in the rhombohedral β-Po geometry near 160 GPa [90]. The β-Po structure can be viewed as a simple cubic (sc) lattice compressed along the body diagonal, and both can be described using the same unit cell with rhombohedral angles of 90° (sc) and 104° (β-Po) [91]. This made us wonder if the metastable *I*4/*mmm*-Na$_2$S$_6$ or *P*4/*mmm*-Na$_2$S$_8$ phases, which both contained sc-like sulfur layers, could potentially be candidates for conventional superconductivity? Calculations were carried out on the latter since it had a higher sulfur content, and a larger normalized density of states at the Fermi level. Figure 10 shows the computed phonon band structure, Eliashberg spectral function, and the electron–phonon integral for *P*4/*mmm*-Na$_2$S$_8$. Because of the similar masses of the sodium and sulfur atoms, both atom types contribute to the phonon modes across the frequency spectrum. The EPC was calculated to be $\lambda = 0.79$ and the logarithmic average frequency $\omega_{\log} = 344.8$ K, resulting in a T_c of 15.5 K using a $\mu^* = 0.1$. For comparison, LDA calculations on the β-Po structure at 200 GPa obtained $\lambda = 0.78$ and $T_c = 19.2$ K using the Allen–Dynes approximation with $\mu^* = 0.11$ [91], and $T_c \sim 16$ K via the multiband approach within the superconducting density functional theory (SCDFT) formalism [92]. Thus, the propensity for superconductivity in *P*4/*mmm*-Na$_2$S$_8$ under pressure appears to be similar to that of elemental sulfur.

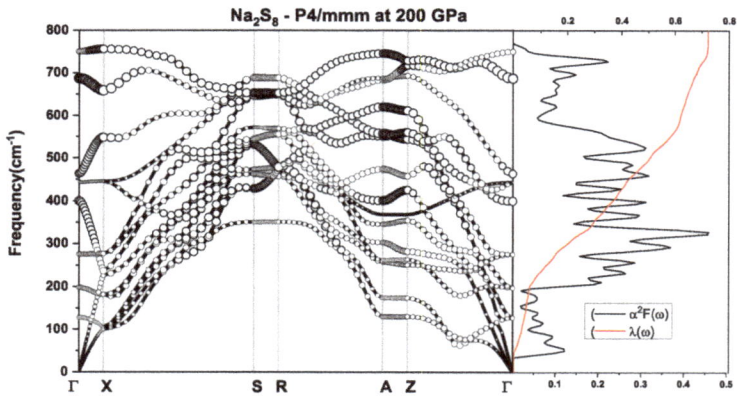

Figure 10. Phonon band structure, Eliashberg spectral function ($\alpha^2 F(\omega)$), and the electron phonon integral ($\lambda(\omega)$) for $P4/mmm$-Na_2S_8 at 200 GPa. Circles indicate the phonon linewidth with a radius proportional to the strength.

4. Conclusions

Evolutionary structure searches coupled with first-principles calculations have been employed to explore the phase diagram of sodium-rich and sodium-poor sulfides at 1 atm and within 100–200 GPa. At ambient pressure, the Na_2S_n, $n = 1, 2, 4, 5$, stoichiometries were thermodynamically stable, whereas $n = 3, 6, 8$ were low lying metastable species. In addition to identifying experimentally known or previously predicted species, we also found the novel $C2/c$-II and $Imm2$-Na_2S_3 phases, which differed in the orientation of the S_3^{2-} anions, as well as $C2$-Na_2S_5, which contained unbranched *trans* S_5^{2-} chains, and $C2$-Na_2S_8, which was comprised of S_8^{2-} helical-like chains. Even though the inclusion of van der Waals interactions did not have much of an impact on the optimized volumes of these phases, in some cases it did have an effect on the computed energy ordering of different polymorphs. Two dynamically stable sodium rich phases, $P\bar{3}m1$-Na_3S and $Pmn2_1$-Na_4S, whose enthalpies of formation lay ∼70 meV/atom above the convex hull were also identified. Their structures were related to $Fm\bar{3}m$-Na_2S, and they had intriguing electronic structures whose metallicity was derived from two-dimensional sodium sheets.

With the exception of the Na_2S stoichiometry, the high-pressure Na-S phase diagram has not been studied before. The stable and metastable sulfur-rich phases contained sulfur motifs that included atoms, dimers, V-shaped trimers, branched polymeric and zigzag chains, as well as two-dimensional corner-linked cyclohexane-like rings, and edge/corner-linked squares. At 200 GPa, the most stable Na_2S_n, $n = 3, 4, 6, 8$, phases were comprised of sodium nets and two-dimensional fused cube or square sulfur nets, whose electronic structure was interrogated. The Allen–Dynes modified McMillan formula was used to predict the superconducting critical temperature, T_c, of Na_2S_8 at 200 GPa, and it was found to be comparable to previous theoretical estimates for the β-Po sulfur phase, 15.5 K for the polysulfide vs. 16–19 K for elemental sulfur [91,92]. At 150–200 GPa, $P6_3/mmc$-Na_3S and $Cmcm$-Na_4S were found to be thermodynamically stable, and their structures differed from those proposed earlier for H_3S [37], Li_3S [35], K_3S [36], Li_4S [35], and K_4S [36] under pressure.

Supplementary Materials: The following are available online at http://www.mdpi.com/2073-4352/9/9/441/s1. Supplementary Materials include the structural coordinates, relative enthalpies, equation of states, volumes, zone-center vibrational frequencies at 0 GPa, electron localization functions, electronic band structures, densities of states, phonon band structures, and electron–phonon coupling parameters for the structures discussed in this paper. Other requests for materials should be addressed to E.Z. (ezurek@buffalo.edu).

Author Contributions: Conceptualization, E.Z.; validation, N.G., T.B., and Y.Y.; formal analysis, N.Z., N.G., T.B., Y.Y., and E.Z.; investigation, N.Z., N.G., T.B., and Y.Y.; resources, E.Z.; writing—original draft preparation, N.G.;

writing—review and editing, E.Z.; visualization, N.G., T.B., and Y.Y.; supervision, E.Z.; project administration, E.Z.; and funding acquisition, E.Z., and Y.Y.

Funding: E.Z. acknowledges the NSF (DMR-1827815) for funding. Support was provided by the Center for Computational Research at the University at Buffalo [93]. Y.Y. acknowledges the National Natural Science Foundation of China Grant No. 11404035, Jilin Provincial Natural Science Foundation of China Grant No. 20190201127JC, and Jilin Province Education Department "13th Five-Year" Science and Technology Research Project Grant No. JJKH20180941KJ.

Conflicts of Interest: The authors declare no conflict of interest.

References

1. Okamoto, H. The Li-S (lithium-sulfur) system. *J. Phase Equilib.* **1995**, *16*, 94–97. [CrossRef]
2. Janz, G.J.; Coutts, J.W.; Downey, J.R.; Roduner, E. Raman studies of sulfur-containing anions in inorganic polysulfides. Potassium polysulfides. *Inorg. Chem.* **1976**, *15*, 1755–1759. [CrossRef]
3. Sangster, J.; Pelton, A.D. The K-S (potassium-sulfur) system. *J. Phase Equilib.* **1997**, *18*, 82–88. [CrossRef]
4. Sangster, J.; Pelton, A.D. The Rb-S (rubidium-sulfur) system. *J. Phase Equilib.* **1997**, *18*, 97–100. [CrossRef]
5. Sangster, J.; Pelton, A.D. The Cs-S (cesium-sulfur) system. *J. Phase Equilib.* **1997**, *18*, 78–81. [CrossRef]
6. Oei, D.G. Sodium-sulfur system. I. Differential thermal analysis. *Inorg. Chem.* **1973**, *12*, 435–437. [CrossRef]
7. Sangster, J.; Pelton, A.D. The Na-S (sodium-sulfur) system. *J. Phase Equilib.* **1977**, *18*, 89–96. [CrossRef]
8. Rosen, E.; Tegman, R. A preparative and X-ray powder diffraction study of the polysulfides Na_2S_2, Na_2S_4 and Na_2S_5. *Acta Chem. Scand.* **1971**, *25*, 3329–3336. [CrossRef]
9. Föppl, H.; Busmann, E.; Frorath, F.K. Die kristallstrukturen von α-Na_2S_2 und K_2S_2, β-Na_2S_2 und Na_2Se_2. *Z. Anorg. Allg. Chem.* **1962**, *314*, 12–20. [CrossRef]
10. Oei, D.G. Sodium-sulfur system. II. Polysulfides of sodium. *Inorg. Chem.* **1973**, *12*, 438–441. [CrossRef]
11. Böttcher, P.; Keller, R. Die kristallstruktur des α-Na_2S_5/The crystal structure of α-Na_2S_5. *Z. Naturforsch. B* **1984**, *39*, 577–581. [CrossRef]
12. Weber, N. Ford gives Na-S battery details. *Chem. Eng. News* **1966**, *44*, 32–33.
13. Kummer, J.T.; Weber, N. Battery Having a Molten Alkali Metal Anode and Molten Sulfur Cathode. U.S. Patent 3,413,150, 26 November 1968.
14. Yao, Y.F.Y.; Kummer, J.T. Ion exchange properties of and rates of ionic diffusion in beta-alumina. *J. Inorg. Nucl. Chem.* **1967**, *29*, 2453–2475.
15. Ellis, B.L.; Nazar, L.F. Sodium and sodium-ion energy storage batteries. *Curr. Opin. Solid State Mater. Sci.* **2012**, *16*, 168–177. [CrossRef]
16. Hueso, K.B.; Armand, M.; Rojo, T. High temperature sodium batteries: Status, challenges and future trends. *Energy Environ. Sci.* **2013**, *6*, 734–749. [CrossRef]
17. Dunn, B.; Kamath, H.; Tarascon, J.M. Electrical energy storage for the grid: A battery of choices. *Science* **2011**, *334*, 928–935. [CrossRef] [PubMed]
18. Nikiforidis, G.; van de Sanden, M.C.M.; Tsampas, M.N. High and intermediate temperature sodium-sulfur batteries for energy storage: Development, challenges and perspectives. *RSC Adv.* **2019**, *9*, 5649–5673. [CrossRef]
19. Momida, H.; Yamashita, T.; Oguchi, T. First-principles study on structural and electronic properties of α-S and Na–S crystals. *J. Phys. Soc. Jpn.* **2014**, *83*, 124713. [CrossRef]
20. Mali, G.; Patel, M.U.M.; Mazaj, M.; Dominko, R. Stable crystalline forms of Na polysulfides: Experiment versus ab initio computational prediction. *Chem. Eur. J.* **2016**, *22*, 3355–3360. [CrossRef]
21. Wang, Y.; Hao, Y.; Xu, L.C.; Yang, Z.; Di, M.Y.; Liu, R.; Li, X. Insight into the discharge products and mechanism of room-temperature sodium-sulfur batteries: A first-principles study. *J. Phys. Chem. C* **2019**, *123*, 3988–3995. [CrossRef]
22. Zintl, E.; Harder, A.; Dauth, B. Gitterstruktur der oxyde, sulfide, selenide und telluride des lithiums, natriums und kaliums. *Z. Elektrochem. Angew. Phys. Chem.* **1934**, *40*, 588–593.
23. May, K. Die kristallstruktur des rubidium-sulfids Rb_2S. *Z. Kristallogr.* **1936**, *94*, 412–413. [CrossRef]
24. Sommer, H.; Hoppe, R. Die Kristallstruktur von Cs_2S. mit einer Bemerkung über Cs_2Se, Cs_2Te, Rb_2Se und Rb_2Te. *Z. Anorg. Allg. Chem.* **1977**, *429*, 118–130. [CrossRef]
25. Grzechnik, A.; Vegas, A.; Syassen, K.; Loa, I.; Hanfland, M.; Jansen, M. Reversible antifluorite to anticotunnite phase transition in Li_2S at high pressures. *J. Solid State Chem.* **2000**, *154*, 603–611. [CrossRef]

26. Vegas, A.; Grzechnik, A.; Syassen, K.; Loa, I.; Hanfland, M.; Jansen, M. Reversible phase transitions in Na$_2$S under pressure: A comparison with the cation array in Na$_2$SO$_4$. *Acta Crystallogr.* **2001**, *B57*, 151–156. [CrossRef]
27. Barkalov, O.I.; Naumov, P.G.; Felser, C.; Medvedev, S.A. Pressure-induced transition to Ni$_2$In-type phase in lithium sulfide(Li$_2$S). *Solid State Sci.* **2016**, *61*, 220–224. [CrossRef]
28. Santamaría-Pérez, D.; Vegas, A.; Muehle, C.; Jansen, M. High-pressure experimental study on Rb$_2$S: Antifluorite to Ni$_2$In-type phase transitions. *Acta Crystallogr.* **2011**, *B67*, 109–115. [CrossRef]
29. Santamaría-Pérez, D.; Vegas, A.; Muehle, C.; Jansen, M. Structural behaviour of alkaline sulfides under compression: High-pressure experimental study on Cs$_2$S. *J. Chem. Phys.* **2011**, *135*, 054511. [CrossRef]
30. Schön, J.C.; Čančarević, Ž.; Jansen, M. Structure prediction of high-pressure phases for alkali metal sulfides. *J. Chem. Phys.* **2004**, *121*, 2289–2304. [CrossRef]
31. Verma, A.K.; Modak, P.; Sharma, S.M. Structural phase transitions in Li$_2$S, Na$_2$S and K$_2$S under compression. *J. Alloy. Compd.* **2017**, *710*, 460–467. [CrossRef]
32. Li, Y.; Jin, X.; Cui, T.; Zhuang, Q.; Lv, Q.; Wu, G.; Meng, X.; Bao, K.; Liu, B.; Zhou, Q. Structural stability and electronic property in K$_2$S under pressure. *RSC Adv.* **2017**, *7*, 7424–7430. [CrossRef]
33. Yao, Y.; Tse, J.S. Superconducting hydrogen sulfide. *Chem. Eur. J.* **2018**, *24*, 1769–1778. [CrossRef]
34. Drozdov, A.P.; Eremets, M.I.; Troyan, I.A.; Ksenofontov, V.; Shylin, S.I. Conventional superconductivity at 203 kelvin at high pressures in the sulfur hydride system. *Nature* **2015**, *525*, 73–76. [CrossRef]
35. Kokail, C.; Heil, C.; Boeri, L. Search for high-T_c conventional superconductivity at megabar pressures in the lithium-sulfur system. *Phys. Rev. B* **2016**, *94*, 060502. [CrossRef]
36. Li, Y.; Jin, X.; Cui, T.; Zhuang, Q.; Zhang, D.; Meng, X.; Bao, K.; Liu, B.; Zhou, Q. Unexpected stable stoichiometries and superconductivity of potassium-rich sulfides. *RSC Adv.* **2017**, *7*, 44884–44889. [CrossRef]
37. Duan, D.; Liu, Y.; Tian, F.; Li, D.; Huang, X.; Zhao, Z.; Yu, H.; Liu, B.; Tian, W.; Cui, T. Pressure-induced metallization of dense (H$_2$S)$_2$H$_2$ with high-T_c superconductivity. *Sci. Rep.* **2014**, *4*, 6968. [CrossRef]
38. Lonie, D.C.; Zurek, E. XtalOpt: An open-source evolutionary algorithm for crystal structure prediction. *Comput. Phys. Commun.* **2011**, *182*, 372–387. [CrossRef]
39. XtalOpt. Available online: http://xtalopt.github.io (accessed on 19 August 2019).
40. Avery, P.; Falls, Z.; Zurek, E. XtalOpt Version r10: An open-source evolutionary algorithm for crystal structure prediction. *Comput. Phys. Commun.* **2017**, *217*, 210–211. [CrossRef]
41. Lonie, D.C.; Zurek, E. Identifying duplicate crystal structures: XtalComp, an open-source solution. *Comput. Phys. Commun.* **2012**, *183*, 690–697. [CrossRef]
42. Avery, P.; Zurek, E. RandSpg: An open-source program for generating atomistic crystal structures with specific spacegroups. *Comput. Phys. Commun.* **2017**, *213*, 208–216. [CrossRef]
43. Kresse, G.; Hafner, J. *Ab initio* molecular dynamics for liquid metals. *Phys. Rev. B* **1993**, *47*, 558–561. [CrossRef]
44. Kresse, G.; Joubert, D. From ultrasoft pseudopotentials to the projector augmented-wave method. *Phys. Rev. B* **1999**, *59*, 1758–1775. [CrossRef]
45. Maintz, S.; Deringer, V.L.; Tchougréeff, A.L.; Dronskowski, R. LOBSTER: A tool to extract chemical bonding from plane-wave based DFT. *J. Comput. Chem.* **2016**, *37*, 1030–1035. [CrossRef]
46. Perdew, J.P.; Burke, K.; Ernzerhof, M. Generalized gradient approximation made simple. *Phys. Rev. Lett.* **1996**, *77*, 3865–3868. [CrossRef]
47. Klimeš, J.; Bowler, D.R.; Michaelides, A. Chemical accuracy for the van der Waals density functional. *J. Phys. Condens. Matter* **2010**, *22*, 022201. [CrossRef]
48. Klimeš, J.; Bowler, D.R.; Michaelides, A. Van der Waals density functionals applied to solids. *Phys. Rev. B* **2011**, *83*, 195131. [CrossRef]
49. Blöchl, P.E. Projector augmented-wave method. *Phys. Rev. B* **1994**, *50*, 17953–17979. [CrossRef]
50. Parlinski, K.; Li, Z.Q.; Kawazoe, Y. First-principles determination of the soft mode in cubic ZrO$_2$. *Phys. Rev. Lett.* **1997**, *78*, 4063–4066. [CrossRef]
51. Chaput, L.; Togo, A.; Tanaka, I.; Hug, G. Phonon-phonon interactions in transition metals. *Phys. Rev. B* **2011**, *84*, 094302. [CrossRef]
52. Togo, A.; Tanaka, I. First principles phonon calculations in materials science. *Scr. Mater.* **2015**, *108*, 1–5. [CrossRef]

53. Giannozzi, P.; Baroni, S.; Bonini, N.; Calandra, M.; Car, R.; Cavazzoni, C.; Ceresoli, D.; Chiarotti, G.L.; Cococcioni, M.; Dabo, I.; et al. QUANTUM ESPRESSO: A modular and open-source software project for quantum simulations of materials. *J. Phys. Condens. Matter* **2009**, *21*, 395502. [CrossRef]
54. Dal Corso, A. Pseudopotentials periodic table: From H to Pu. *Comput. Mater. Sci.* **2014**, *95*, 337–350. [CrossRef]
55. Troullier, N.; Martins, J.L. Efficient pseudopotentials for plane-wave calculations. *Phys. Rev. B* **1991**, *43*, 1993–2006. [CrossRef]
56. Methfessel, M.; Paxton, A.T. High-precision sampling for Brillouin-zone integration in metals. *Phys. Rev. B* **1989**, *40*, 3616–3621. [CrossRef]
57. Allen, P.B.; Dynes, R.C. Transition temperature of strong-coupled superconductors reanalyzed. *Phys. Rev. B* **1975**, *12*, 905–922. [CrossRef]
58. Hanfland, M.; Loa, I.; Syassen, K. Sodium under pressure: Bcc to fcc structural transition and pressure-volume relation to 100 GPa. *Phys. Rev. B* **2002**, *65*, 184109. [CrossRef]
59. Gregoryanz, E.; Lundegaard, L.F.; McMahon, M.I.; Guillaume, C.; Nelmes, R.J.; Mezouar, M. Structural diversity of sodium. *Science* **2008**, *320*, 1054–1057. [CrossRef]
60. McMahon, M.I.; Gregoryanz, E.; Lundegaard, L.F.; Loa, I.; Guillaume, C.; Nelmes, R.J.; Kleppe, A.K.; Amboage, M.; Wilhelm, H.; Jephcoat, A.P. Structure of sodium above 100 GPa by single-crystal x-ray diffraction. *Proc. Natl. Acad. Sci. U. S. A.* **2007**, *104*, 17297–17299. [CrossRef]
61. Ma, Y.; Eremets, M.; Oganov, A.R.; Xie, Y.; Trojan, I.; Medvedev, S.; Lyakhov, A.O.; Valle, M.; Prakapenka, V. Transparent dense sodium. *Nature* **2009**, *458*, 182–185. [CrossRef]
62. Meyer, B. Elemental sulfur. *Chem. Rev.* **1976**, *76*, 367–388. [CrossRef]
63. Zakharov, O.; Cohen, M.L. Theory of structural, electronic, vibrational, and superconducting properties of high-pressure phases of sulfur. *Phys. Rev. B* **1995**, *52*, 12572–12578. [CrossRef]
64. Gavryushkin, P.N.; Litasov, K.D.; Dobrosmislov, S.S.; Popov, Z.I. High-pressure phases of sulfur: Topological analysis and crystal structure prediction. *Phys. Status Solidi B* **2017**, *254*, 1600857. [CrossRef]
65. Luo, H.; Greene, R.G.; Ruoff, A.L. β-Po phase of sulfur at 162 GPa: X-ray diffraction study to 212 GPa. *Phys. Rev. Lett.* **1993**, *71*, 2943–2946. [CrossRef]
66. Grimme, S. Semiempirical GGA-type density functional constructed with a long-range dispersion correction. *J. Comput. Chem.* **2006**, *27*, 1787–1799. [CrossRef]
67. Lozano, A.; Escribano, B.; Akhmatskaya, E.; Carrasco, J. Assessment of van der Waals inclusive density functional theory methods for layered electroactive materials. *Phys. Chem. Chem. Phys.* **2017**, *19*, 10133–10139. [CrossRef]
68. Perdew, J.P.; Ruzsinszky, A.; Csonka, G.I.; Vydrov, O.A.; Scuseria, G.E.; Constantin, L.A.; Zhou, X.; Burke, K. Restoring the density-gradient expansion for exchange in solids and surfaces. *Phys. Rev. Lett.* **2008**, *100*, 136406. [CrossRef]
69. Constantin, L.A.; Terentjevs, A.; Della Sala, F.; Fabiano, E. Gradient-dependent upper bound for the exchange-correlation energy and application to density functional theory. *Phys. Rev. B* **2015**, *91*, 041120. [CrossRef]
70. Peng, H.; Perdew, J.P. Rehabilitation of the Perdew-Burke-Ernzerhof generalized gradient approximation for layered materials. *Phys. Rev. B* **2017**, *95*, 081105. [CrossRef]
71. Terentjev, A.V.; Constantin, L.A.; Pitarke, J. Dispersion-corrected PBEsol exchange-correlation functional. *Phys. Rev. B* **2018**, *98*, 214108. [CrossRef]
72. Peng, H.; Yang, Z.H.; Perdew, J.P.; Sun, J. Versatile van der Waals density functional based on a meta-generalized gradient approximation. *Phys. Rev. X* **2016**, *6*, 041005. [CrossRef]
73. Rettig, S.J.; Trotter, J. Refinement of the structure of orthorhombic sulfur, α-S_8. *Acta Cryst.* **1987**, *C43*, 2260–2262. [CrossRef]
74. Wagman, D.D.; Evans, W.H.; Parker, V.B.; Schumm, R.H.; Halow, I.; Bailey, S.M.; Churney, K.L.; Nuttall, R.L. The NBS tables of chemical thermodynamic properties. *J. Phys. Chem. Ref. Data* **1982**, *11* (Suppl. 2), 1–392.
75. Tegman, R. The crystal structure of sodium tetrasulphide, Na_2S_4. *Acta Cryst.* **1973**, *B29*, 1463–1469. [CrossRef]
76. Kelly, B.; Woodward, P. Crystal structure of dipotassium pentasulphide. *J. Chem. Soc. Dalton Trans.* **1976**, 1314–1316. [CrossRef]

77. Böttcher, P. Synthesis and crystal structure of the dirubidiumpentachalcogenides Rb_2S_5 and Rb_2Se_5. *Z. Kristallogr. Cryst. Mater.* **1979**, *150*, 65–73. [CrossRef]

78. Böttcher, P.; Kruse, K. Darstellung und kristallstruktur von dicaesiumpentasulfid (Cs_2S_5). *J. Less Common Met.* **1982**, *83*, 115–125. [CrossRef]

79. Janz, G.J.; Downey, J.R.; Roduner, E.; Wasilczyk, G.J.; Coutts, J.W.; Eluard, A. Raman studies of sulfur-containing anions in inorganic polysulfides. Sodium polysulfides. *Inorg. Chem.* **1976**, *15*, 1759–1763. [CrossRef]

80. Te Velde, G.; Bickelhaupt, F.M.; Baerends, E.J.; Fonseca Guerra, C.; van Gisbergen, S.J.A.; Snijders, J.G.; Ziegler, T. Chemistry with ADF. *J. Comput. Chem.* **2001**, *22*, 931–967. [CrossRef]

81. ADF2018. SCM, Theoretical Chemistry, Vrije Universiteit, Amsterdam, The Netherlands. Available online: http://www.scm.com (accessed on 19 August 2019).

82. Feng, J.; Hennig, R.G.; Ashcroft, N.; Hoffmann, R. Emergent reduction of electronic state dimensionality in dense ordered Li-Be alloys. *Nature* **2008**, *451*, 445–458. [CrossRef]

83. Miao, M. Helium chemistry: React with nobility. *Nat. Chem.* **2017**, *9*, 409–410. [CrossRef]

84. Liu, Z.; Botana, J.; Hermann, A.; Zurek, E.; Yan, D.; Lin, H.; Miao, M. Reactivity of He with ionic compounds under high pressure. *Nat. Commun.* **2018**, *9*, 951. [CrossRef]

85. Zarifi, N.; Liu, H.; Tse, J.S.; Zurek, E. Crystal structures and electronic properties of Xe-Cl compounds at high pressure. *J. Phys. Chem. C* **2018**, *122*, 2941–2950. [CrossRef]

86. Mishra, A.K.; Muramatsu, T.; Liu, H.; Geballe, Z.M.; Somayazulu, M.; Ahart, M.; Baldini, M.; Meng, Y.; Zurek, E.; Hemley, R.J. New calcium hydrides with mixed atomic and molecular hydrogen. *J. Phys. Chem. C* **2018**, *122*, 19370–19378. [CrossRef]

87. Gregoryanz, E.; Degtyareva, O.; Somayazulu, M.; Hemley, R.J.; Mao, H.K. Melting of dense sodium. *Phys. Rev. Lett.* **2005**, *94*, 185502. [CrossRef]

88. Neaton, J.B.; Ashcroft, N.W. On the constitution of sodium at higher densities. *Phys. Rev. Lett.* **2001**, *86*, 2830–2833. [CrossRef]

89. Tutchton, R.; Chen, X.; Wu, Z. Is sodium a superconductor under high pressure? *J. Chem. Phys.* **2017**, *146*, 014705. [CrossRef]

90. Struzhkin, V.V.; Hemley, R.J.; Mao, H.K.; Timofeev, Y.A. Superconductivity at 10–17 K in compressed sulphur. *Nature* **1997**, *390*, 382–384. [CrossRef]

91. Rudin, S.P.; Liu, A.Y. Predicted simple-cubic phase and superconducting properties for compressed sulfur. *Phys. Rev. Lett.* **1999**, *83*, 3049–3052. [CrossRef]

92. Monni, M.; Bernardini, F.; Sanna, A.; Profeta, G.; Massidda, S. Origin of the critical temperature discontinuity in superconducting sulfur under high pressure. *Phys. Rev. B* **2017**, *95*, 064516. [CrossRef]

93. Center for Computational Research, University at Buffalo. Available online: http://hdl.handle.net/10477/79221 (accessed on 19 August 2019).

crystals

MDPI

Article

Simultaneous Prediction of the Magnetic and Crystal Structure of Materials Using a Genetic Algorithm

Edward J. Higgins, Phil J. Hasnip and Matt I.J. Probert *

Department of Physics, University of York, Heslington, York YO10 5DD, UK
* Correspondence: matt.probert@york.ac.uk; Tel.: +44-1904-322239

Received: 15 July 2019; Accepted: 16 August 2019; Published: 23 August 2019

Abstract: We introduce a number of extensions and enhancements to a genetic algorithm for crystal structure prediction, to make it suitable to study magnetic systems. The coupling between magnetic properties and crystal structure means that it is essential to take a holistic approach, and we present for the first time, a genetic algorithm that performs a simultaneous global optimisation of both magnetic structure and crystal structure. We first illustrate the power of this approach on a novel test system—the magnetic Lennard–Jones potential—which we define. Then we study the complex interface structures found at the junction of a Heusler alloy and a semiconductor substrate as found in a proposed spintronic device and show the impact of the magnetic interface structure on the device performance.

Keywords: structure prediction; magnetic materials; genetic algorithm; global optimisation; ab initio; DFT; structural fingerprint; magnetic Lennard–Jones; Heusler alloy; half-Heusler alloy

---

1. Introduction

In order to meet the challenges posed by modern and emerging technologies, it is increasingly necessary to look beyond existing, known materials. Many fields, from solar cells to spintronic devices, call for materials with unprecedented performance characteristics, or even entirely new behavior. Searching for new materials experimentally is expensive and time-consuming, but the advent of efficient, accurate computational materials modeling offers a potential way forward. Magnetic materials are of particular interest, with applications from fast, high-density data storage such as magnetic RAM devices [1] and heat-assisted magnetic recording (HAMR) [2], to new spintronic and quantum devices, such as spin-valves [3,4]. Magnetic materials include conventional ferromagnets along with more exotic structures, such as antiferromagnets, ferrimagnets, and spin glasses. These materials are already at the heart of many important technologies, but play an increasingly important role in developing and future technologies.

Many of the strong permanent magnets in use today rely on rare-earth elements [5] and there is concern over the sustainability of these elements' availability [6]. Therefore, developing ferromagnets made from more readily available materials is desirable.

In order to determine a material's properties, it is not sufficient to know its chemical composition, it is also vital to understand the crystal structure of the material. In general, atoms will arrange themselves in a material so as to minimise the total (free) energy. One method to determine crystal structure is to estimate an initial atomic configuration, compute the atomic forces and lattice cell stress according to a suitable model, and then adjust the atomic positions and lattice parameters in order to minimise the energy (where the forces and stresses are zero). This procedure works well if the initial geometry is sufficiently close to the true ground state, and this workflow has been the backbone of computational materials studies for many decades. However, the resultant optimised structure only represents the local energy minimum; that is, this procedure finds the lowest energy structure which is closest to the initial geometry. It is possible that a lower-energy structure exists, but that it is separated from the initial geometry by an energy barrier which the local optimisation procedure will not overcome. This situation is depicted in Figure 1.

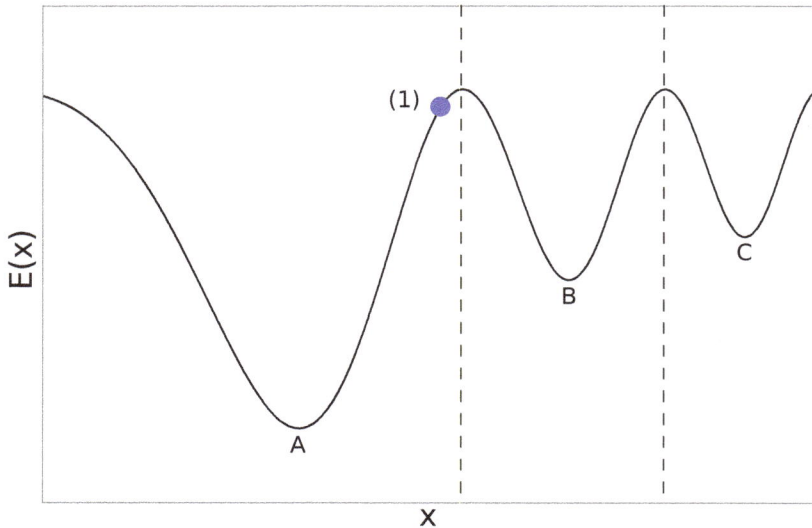

Figure 1. An example of global (A) and local (B, C) minima for a function $f(x)$. The dashed lines mark the boundaries between the different basins of attraction. A local minimisation method will find a different minimum (A, B or C) depending upon which basin of attraction the starting point (1) is in. Note: point (1) is closer in x to B, but is in the basin of attraction of A.

In fact, many materials do possess a range of low-energy phases, and determining the stable phase(s) under given environmental conditions is extremely challenging, even for well-studied classes of materials. In general, finding the full range of stable phases requires a global optimisation method, usually involving a wide range of simulations, encompassing atomic configurations which span all the possible phases. Developing global optimisation methods is an active research field in materials modelling, and existing methods include basin hopping [7–9], random structure searching [10], meta-dynamics [11], swarm optimisation [12] and evolutionary algorithms. The latter class includes genetic algorithms [13–16], which have met with great success in materials modelling and are the focus of this work.

Conventional global optimisation methods for crystal structure prediction often work in the basis of the atomic positions and lattice vectors. However when considering systems with nontrivial magnetic structures, this representation leads to a multi-valued energy landscape to search, even when the crystal structure itself is relatively well-established. As a result of this, conventional algorithms can struggle when it comes to searching the combined phase space of magnetic- and crystal-structures for magnetic materials.

Therefore, in this work, we present a novel genetic algorithm (GA) for the simultaneous global optimisation of both the magnetic structure and the crystal structure of a material. We shall explain the key ideas, and then illustrate with two example studies. The first is a new model system, suitable for the study of magnetic systems at the atomic scale—the magnetic Lennard-Jones potential. Whilst the conventional Lennard–Jones potential is a well-known test system for many structural and dynamical algorithms, it has not yet (to our knowledge) been extended to study magnetic systems. We shall introduce this system and explore some of its fundamental behaviour as a means to test our new magnetic GA. The second system we shall study is more complex, and is inspired by recent experiments on spintronic systems, and is an interface between a Heusler alloy and a Ge substrate. We shall use the magnetic GA to find the optimal interface structure and show how the novel predicted structure can explain the experimental data.

2. Materials and Methods

2.1. Materials Modelling

The first step in finding a stable crystal structure is to have a reasonable model for the energy of any particular atomic configuration. There are two main classes of models in this context: classical forcefields; and quantum mechanical methods.

Classical forcefields are multivariate functions which take the atomic species and positions as input, and return the internal energy of the system and the atomic forces (the first derivative of the energy with respect to the atomic positions). Since atomic interactions depend on the positions of the atoms relative to each other, rather than the absolute positions themselves, forcefields are usually expressed in terms of bond lengths and, depending on the particular forcefield, bond angles and bond torsions. One of the earliest and simplest interatomic forcefields was proposed by Lennard–Jones [17], and expresses the energy E^{LJ} of an atomic configuration as

$$E^{LJ} = \sum_{i,j \neq i} 4\epsilon_{ij} \left[\left(\frac{\sigma_{ij}}{r_{ij}} \right)^{12} - \left(\frac{\sigma_{ij}}{r_{ij}} \right)^{6} \right], \tag{1}$$

where r_{ij} is the distance between atoms i and j and ϵ_{ij} and σ_{ij} are parameters of the potential which control the equilibrium bond energy and bond-length, respectively.

Forcefields are generally computationally cheap and simulations of millions of atoms may be carried out routinely. The main drawbacks of forcefield methods are the need to select and parameterise an appropriate forcefield for the material, and the inability of most forcefields to model dynamic changes in chemistry such as bond breaking or formation.

Quantum mechanical approaches centre on solving the many-body Schrödinger equation for the electrons and nuclei which comprise the material. These approaches have two principal advantages over forcefield methods: they are parameter-free since the form of the Schrödinger equation is known exactly; and by modeling the electrons explicitly, the approaches handle seamlessly chemical complexities such as bond formation and breaking, or environmentally-dependent oxidation states. There is one significant drawback, however, in that this is an extremely computationally intensive task and solving the many-body Schrödinger equation in its original form is unfeasible for materials. Most practical quantum-mechanical materials simulations are based on a reformulation of quantum mechanics known as density functional theory (DFT).

DFT is founded on the Hohenberg-Kohn theorem [18], which proves that the ground state energy of a many-body quantum system is a unique, universal functional of the particle density and has no explicit dependence on the many-body wavefunction. Kohn and Sham [19] used this to demonstrate a formal link between the ground state of the many-body Schrödinger equation, and the ground state of a related auxiliary set of coupled single-particle equations. Crucially, these auxiliary equations may be solved with far less computational resources than the original many-body expressions [20,21].

In principle the mapping between the many-body system and the auxiliary system is exact, but there is one term in the auxiliary equations which is unknown: the exchange-correlation potential. The exchange-correlation potential must be approximated in practical calculations, with the most common approximations being built on known limits and numerically-exact results for weakly interacting electron gases, e.g., the approximation of Perdew, Burke and Ernzerhof [22]. Despite these approximations, DFT has enjoyed popularity and great success in a wide range of materials simulations [23,24].

2.2. Genetic Algorithms

A common solution to the need for global energy optimisation is to use an ensemble, or 'population', of starting atomic geometries. Each of these candidate geometries can then be optimised independently with a local optimisation method, in an attempt to map out the possible local minima. Since the initial geometries are unlikely to span enough of the configuration space to find all the local minima, the geometries are typically updated in order to explore other regions of the configuration space. Due to the independent nature of these members, population-based methods often have more scope for parallelism than other methods.

One popular class of population-based methods for structure prediction is that of genetic algorithms (GAs) [13–16]. In these algorithms, the 'fitness' of members of a population is evaluated in some appropriate manner (e.g., from the binding energy per atom) in an attempt to decide which of them are most likely to find solutions of interest. The population is then updated by selecting favorable (e.g., low-energy) members of the population, and generating new population members by combining these members to produce 'child' members, a process known as 'crossover', followed by some random mutation(s). Depending on the problem at hand, a variety of methods exist for this fitness evaluation, member recombination, and selection stages.

When considering how to represent an optimisation problem for a genetic algorithm, a choice needs to be made about how the search vector is to be represented for these operations. A simple GA may treat the input vector simply as a 1D array of real numbers or as a bit string representing that vector. However, it is often the case that a more physically meaningful representation of the search vector can improve the convergence characteristics of the algorithm. In addition to pure GAs, a number of hybrid GAs exist, incorporating ideas from other algorithms. For example, memetic algorithms allow each member of the population to perform local searches.

In this work, we started from the GA developed by Abraham and Probert [14], which was created specifically for the prediction of crystal structures in periodic boundary conditions. In this GA, the crossover is performed in real-space by using a pair of periodic cuts to select material from each parent structure. This material is combined to form child structures, as illustrated in Figure 2.

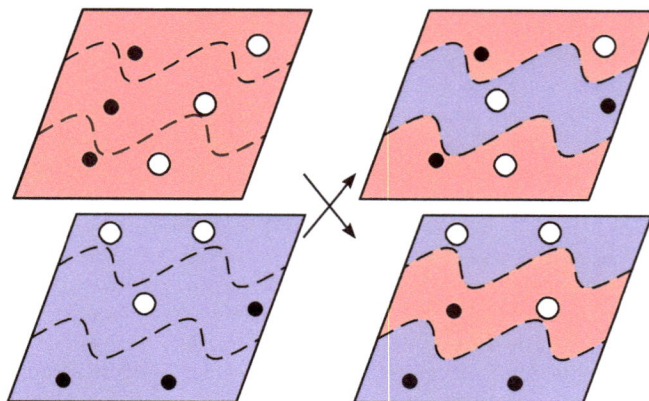

Figure 2. Real-space crossover of two unit cells, each containing six atoms across two different atomic elements (black and white circles) in two dimensions using a pair of periodic cuts (dashed lines). The cuts define two sets of atoms for each of the parents (**left**): the 'inner' set comprises atoms lying between the two cuts; and the 'outer' set comprises the remainder of the atoms. Child structures (**right**) are formed by combining atoms from the 'inner' set of one parent, with those from the 'outer' set of the other parent.

Following crossover, the child structures are mutated before a local optimisation method is performed. Several different types of mutations are allowed: deformation of the cell vectors; perturbations of the atomic positions; and inter-atomic swaps (see Figure 3). Mutations are important to GAs, as they are the only way of introducing fundamentally new structural features which are not present in previous populations.

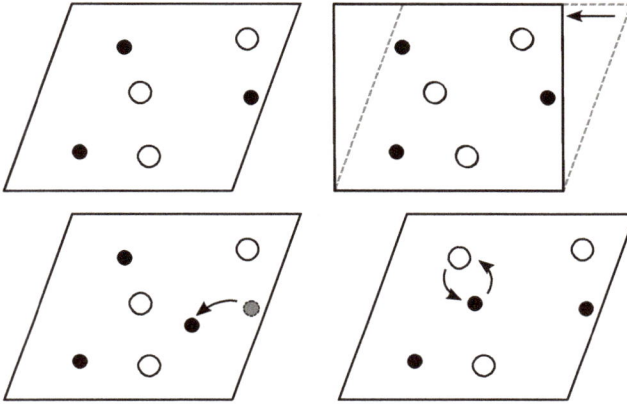

Figure 3. Example of mutation operations on one of the child structures from Figure 2. The original cell (**top left**) can be mutated by: changing the cell vectors (**top right**); perturbing the atomic positions (**bottom left**); or permuting the atoms (**bottom right**).

Once the local optimisation method has optimised all of the population structures, the fitness of the structures is evaluated. Based on the fitness, some structures are removed from the population (preferentially those with low fitness) and some structures are chosen for crossover (preferentially those with high fitness).

One potential weakness of the method outlined thus far is the tendency of population members to become more and more similar, a process known as 'stagnation'. In extreme cases, every single population member may converge to the same structure; this structure usually represents a stable, low-energy phase but it is not necessarily the global minimum-energy structure, and once a population has stagnated it is almost impossible for it to diversify again significantly.

In order to prevent stagnation, the GA was extended in Ref. [25,26] to incorporate a structure-factor based fingerprinting technique, which enables the GA to differentiate structures in order to penalise structures that are too similar, lowering their fitness and encouraging diversity within the population. For a given structure x containing N atoms, the (non-magnetic) fingerprint is defined as:

$$\Lambda(k) = \sum_{i=1}^{N} \rho_i^2 + 2 \sum_{i=1}^{N} \sum_{j>i}^{N} \rho_i \rho_j \times j_0\left(k|\vec{r}_i - \vec{r}_j|\right), \tag{2}$$

where $\vec{r}_i$ is the position of atom i, ρ_i is the charge density of the nucleus of atom i, defined to be Z_i at $\vec{r}_i$ and zero elsewhere, and j_0 is the spherical Bessel function of the first kind. For each structure, the value of Λ is calculated for a range of values of wavenumber k.

Using this fingerprint, the difference between two structures can then be quantified, for example using an R-factor inspired by the Pendry R-factor. Using this, the difference between two structures x and x' is defined as:

$$R_{xx'} = \frac{\sum_k |\Lambda_x(k) - \Lambda_{x'}(k)|}{\sum_k \Lambda_x(k)}. \tag{3}$$

2.3. Extending the GA for Magnetic Materials

A GA may be extended to optimise magnetic systems by the simple expedient of including magnetic effects in the calculation of energies and forces, and allowing the local optimisation method to minimise the energy with respect to the magnetic degrees of freedom as well as the crystal structure. However, this naïve approach suffers from two severe problems: firstly, the magnetic energy landscape itself has multiple minima, necessitating a global energy optimisation method; secondly, the magnetic structure and crystal structure of a material are often coupled, and may not be treated independently of each other. Elemental iron may be considered a prototypical example of both effects, possessing an antiferromagnetic face-centred cubic phase in addition to the familiar ferromagnetic body-centred cubic phase (see Figure 4). For these reasons it is important to consider the magnetic- and atomic-structure of a material on an equal footing, and to be able to predict both simultaneously.

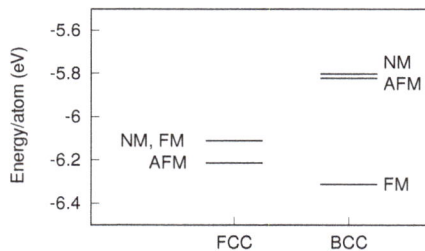

Figure 4. Comparison of the binding energies per atom of iron in the face-centred cubic (FCC) and body-centred cubic (BCC) structures, for ferromagnetic (FM), antiferromagnetic (AFM) and non-magnetic (NM) arrangements of the atomic spins. This illustrates the importance of both crystal structure and magnetic ordering on the overall stability of a magnetic system. Values from [27].

In order to account for magnetic effects within the GA itself, the magnetism needs to be included within the representation of the structure and evolved using updated GA operations such as crossover and mutations. For the representation of magnetism in the system, this work assigns an atomic spin to each atom, which can be either an additional degree of freedom for collinear spin systems or an additional three degrees of freedom for non-collinear spin systems. When considering DFT simulations, spin information will typically be represented as the electronic spin density across all space. In order to project this onto the atomic representation, Hirshfeld analysis [28] was used to partition the electronic spin density into regions of space associated with each atom. This information can then be transmitted from parent structures to offspring during the crossover operation.

2.3.1. Perturbation/Permutation Operations

In order for a GA to optimise the spins efficiently, mutation operations need to be extended to also affect the spin degrees of freedom. This work defines two spin mutations: perturbation and permutation of the atomic spins.

In the case of perturbations, a distinction needs to be made between collinear and non-collinear spin systems. For collinear spin systems, there is only one scalar value per atom to optimise. The perturbation therefore takes the same role as the atomic position perturbations, where the role of the perturbation is to move the system from one basin of attraction to another. Since there is a maximum value these spins can take, i.e., the total number of electrons associated with the atom, issues may arise trying to add an additional perturbation to an existing spin. For example, an atom with saturated spin given a positive perturbation will become unphysical. As a result, the spin perturbation is evaluated as a uniform random number between -spin_max and spin_max, where spin_max is a user parameter, set to the maximum spin value expected on any atom. For example, for a d-block transition metal, spin_max should be set to 5 $\hbar/2$, since the d shell is capable of having a maximum spin of $5\,\hbar/2$. This perturbation allows atomic spins to spontaneously magnetise, demagnetise or flip, irrespective of the initial value.

In the case of non-collinear spins, a similar procedure is performed. In this case, the atomic spin is set to a random vector within the sphere of radius spin_max. Again, this allows any spin state to be found by the perturbation within the range specified, without the risk of being stuck in a state based on the structure's history.

For permutations, the situation is analogous to swapping the atomic positions of different species. However, since atomic spin is closely related to the crystal structure, it only makes sense to swap atomic spins of atoms of the same species. For example, swapping the spin of a nickel and oxygen atom in NiO makes no physical sense, since all the spin exists on the Ni atoms, and the O atoms generally have zero spin.

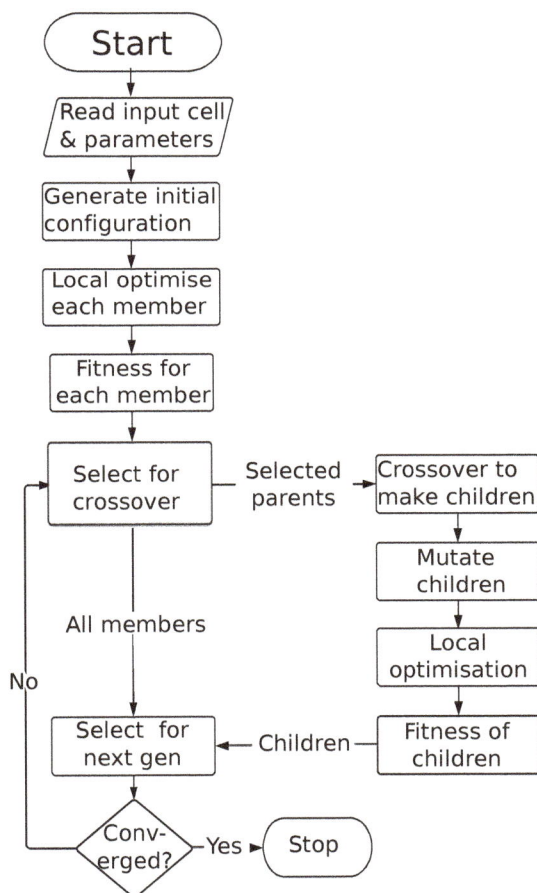

Figure 5. Schematic program flow of a genetic algorithm, as implemented in the CASTEP materials modelling package.

We shall perform all our calculations using the general purpose DFT code CASTEP [29], which relaxes the electronic charge density and spin (in either collinear or non-collinear form) to find the electronic ground state. There may be multiple local minima with different spin configurations.

Hence the input atomic spin from the GA is used to initialise the electron density, and CASTEP then acts as a local optimiser for the atomic spin. This process may then be repeated as CASTEP performs a geometry optimisation to find the (local) minimum energy configuration of atomic positions and cell vectors (Figure 5).

2.3.2. Magnetic Fingerprinting

To encourage diversity in the population, it is important to be able to quantify the differences between structures. In Abraham and Probert [25,26], a fingerprint was introduced that was translationally and rotationally invariant, and could successfully distinguish different structures. However, it will fail to distinguish two magnetic systems that have the same crystal structure but a different arrangement of spins. Hence we propose an augmented fingerprint Λ_{aug}, inspired by magnetic scattering experiments, that can differentiate magnetic structures [30]:

$$\Lambda_{\mathrm{aug}}(k) = \Lambda(k) + q^2 \Lambda_{\mathrm{magn}}(k) \tag{4}$$

where $\Lambda(k)$ is the original crystal structure fingerprint as defined in Equation (2) and

$$\Lambda_{\mathrm{magn}}(k) = \sum_{i=1}^{N} S_i^2 + 2 \sum_{i=1}^{N} \sum_{j>i}^{N} S_i S_j \times j_0 \left(k |\vec{r}_i - \vec{r}_j| \right) \tag{5}$$

is the magnetic structure fingerprint term. Here S_i is the spin on atom i and q is some scaling parameter which allows the overall fingerprint to be more or less sensitive with respect to differences in magnetic structure and crystal structure.

The R-factor in Equation (3) can be used with this augmented fingerprint to quantify the difference between two magnetic structures. As an example, the effect of perturbations on a 6-atom Fe unit cell is shown in Figure 6. The black crosses represent the R-factor difference of structure with perturbed atomic positions with respect to the original reference structure. The red squares and blue circles represent similar perturbations with an additional perturbation to the atomic spins. Here a value of $q = 5$ was used to make significantly different spin structures ($\Delta S \geq \hbar/2$) appear as different as significantly different crystal structures ($\Delta r > 0.2$ Å), both of which can be represented by an R-factor difference of $R_{xx'} > 0.03$.

2.4. Case Studies

2.4.1. Fictional Magnetic Potential: LJ + S

In order to test the GA on magnetic systems, a pair-potential model including magnetic effects was defined. Since a significant portion of these effects are intrinsically quantum mechanical or many-body in nature, it is difficult to capture all of these effects in empirical potentials without limiting them to some specific regime. Magnetic moments on isolated atoms tend to arise from partially filled electronic states, as dictated by Hund's rules. As a result, atoms with nonzero magnetic moments tend to be polyvalent, causing an additional challenge to empirical pair potentials.

Since magnetic effects are often too complex to include explicitly in pair-potentials, their effects are usually included implicitly through the mechanical and thermal properties used to parameterise them [31]. There are some empirical potentials however that try to explicitly include magnetic effects [31–33] and some machine learning models have been trained to deal with magnetic systems [34]. These models tend to require far more complex effects than simply pair-wise interactions.

Given all this, attempting to describe real magnetic materials using a pair-potential is beyond the scope of this work. Instead, the spatial dependence of the energy from the standard (non-magnetic) Lennard-Jones potential was combined with the magnetic dependence of the energy from the lattice-based Heisenberg model in order to create a potential which behaves somewhat like a magnetic material. This work defines a generalised magnetic Lennard-Jones potential as:

$$V_{ij}^{LJ+S} = V_{ij}^{LJ} - \underbrace{A\vec{S}_i \cdot \vec{S}_j f_{\text{ex}}(\vec{R}_{ij})}_{\text{Exchange term}} + \underbrace{\frac{B}{|\vec{R}_{ij}|^3}\left[\vec{S}_i \cdot \vec{S}_j - \frac{3}{|\vec{R}_{ij}^2|}(\vec{S}_i \cdot \vec{R}_{ij})(\vec{S}_j \cdot \vec{R}_{ij})\right]}_{\text{Dipole term}} + \underbrace{C f_{\text{ani}}(\vec{R}_{ij}, \vec{S}_i, \vec{S}_j)}_{\text{Anisotropy term}} \quad (6)$$

In general, a magnetic material may have contributions to the potential from exchange between sites i and j (with spatial dependence given by the $f_{ex}(r_{ij})$ term; a dipole-dipole interaction of the given form; and a symmetry-dependent anisotropy term.

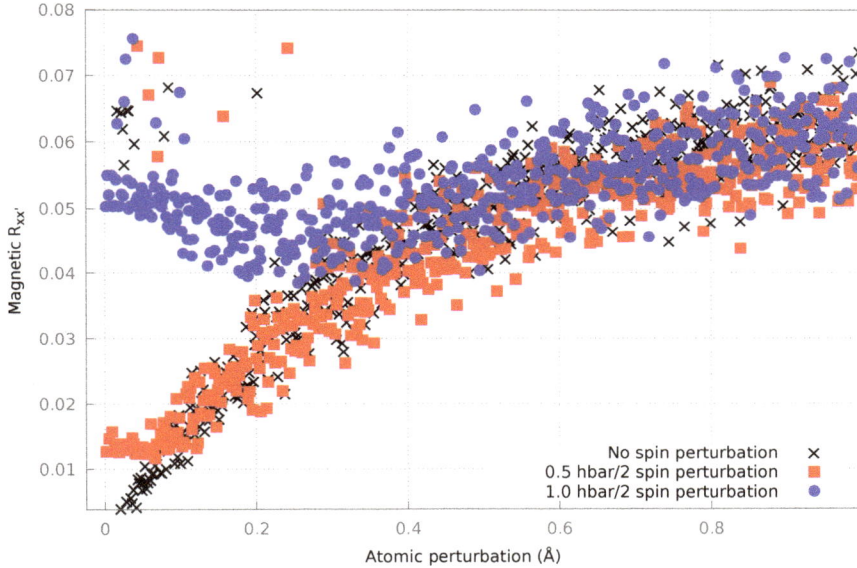

Figure 6. R-factor difference (Equation (3)) comparing the magnetic fingerprints (Equation (4)) of a 6-atom Fe unit cell to the same cell with a range of perturbations applied to the atomic positions. In addition to just performing perturbations to the positions (black crosses), the processes was repeated with an additional perturbation to the atomic spins (red squares and blue circles).

In this study, the magnetic GA was tested on a reduced form of this potential where $B = C = 0$ and $f_{ex} = \exp(-\alpha r_{ij})$. However this form of f_{ex} could be modified to include more complex forms of the exchange interaction such as the RKKY interaction. The form of the potential used is shown in Figure 7.

2.4.2. CFAS/n-Ge Interface

One material that is of interest for its magnetic properties is the full Heusler alloy $Co_2FeAl_{0.5}Si_{0.5}$ (CFAS). CFAS is a half-metal and has been proposed as a candidate material for novel spintronic applications such as spin valves and magnetic tunnel junctions. In order to use Heusler alloys for device applications, the material needs to be attached to electrical contacts. For a number of spintronic applications of Heusler alloys, both metallic and semiconducting contacts are required.

It is known that the half-metallic nature of Heusler alloys at the interface can be sensitive to the exact configuration of the interface atoms, and this can affect the performance of the device. For example, it is known that Co_2MnSi (CMS) grown on a silver surface changes significantly depending on whether it terminates with a Co layer or a Mn/Si layer [35].

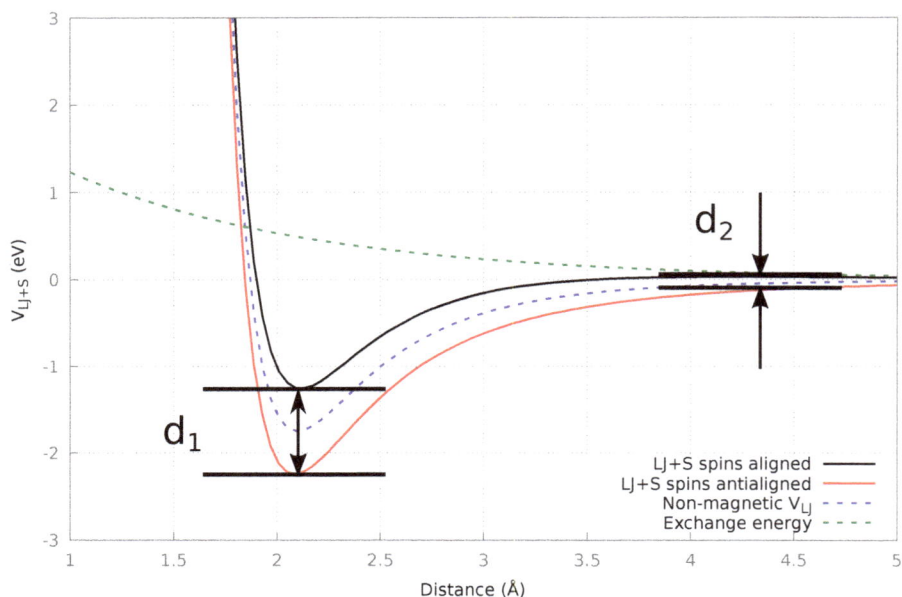

Figure 7. Analytic form of the LJ+S potential, showing the V^{LJ+S} for aligned and anti-aligned spins (solid black/red respectively), along with the contributions from the V^{LJ} and magnetic terms (dashed blue/green respectively). The difference in energy between aligned and antialigned pairs at r_{min} and $2r_{min}$ (d_1 and d_2 respectively) were derived from DFT calculations on an Fe dimer to parameterise the potential.

For the CFAS/semiconductor interface, germanium provides an extremely good lattice match with only a 0.2% mismatch, compared to around 4.5% for CFAS/Si. However, when CFAS is grown on n-type germanium, significant mixing at the interface can occur, as evidenced by energy dispersive X-ray spectroscopy (EDS) [36] (see Figure 8).

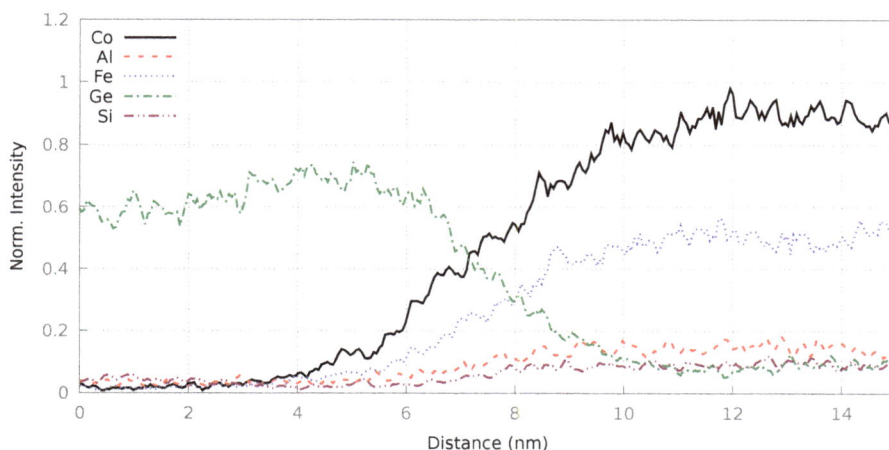

Figure 8. Energy dispersive X-ray spectroscopy (EDS) results across the as-grown CFAS/n-Ge interface, showing significant mixing of the atomic species. Data taken from Kuerbanjian et al. [36].

In addition, annealing the sample at 575 K forms a plateau in the cobalt intensities and, to a lesser extent, the iron and germanium intensities, as seen in the EDS results in Figure 9. This suggests that a stable phase is forming at this point. Since this is in the centre of the interface region and all the EDS intensities are about half of their respective bulk values, it is proposed that this structure contains half the number of atoms in the bulk CFAS and half the number of atoms in a germanium unit cell. This results in a potential $Ge_4Co_4Fe_2AlSi$ structure, containing 12 atoms in the formula unit, although the crystal structure of this interface phase is unknown. Since the electronic properties of this phase will affect the half-metallicity of the CFAS/n-Ge interface, the structure of this phase was investigated using the new magnetic GA.

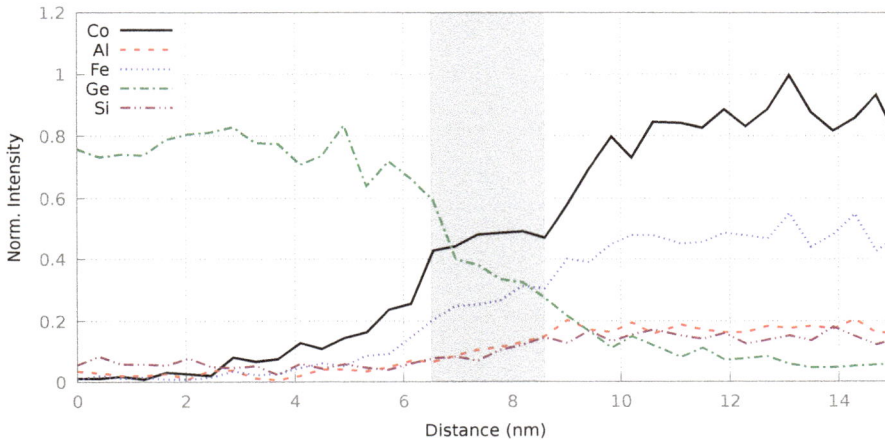

Figure 9. EDS results across the annealed CFAS/n-Ge interface, showing significant mixing of the atomic species. Plateaux can be seen around 6.5–8.6 nm in Co, Fe and Ge, indicated by the shaded region. Data taken from Kuerbanjian et al. [36].

3. Results and Discussion

3.1. LJ+S

For real-world magnetic materials, it is often the exchange term which dominates the magnetic interaction. Because of this, the work presented in this section will use a simplified version of the LJ+S potential, ignoring the dipole and anisotropy terms (i.e., choosing $B = C = 0$) and using the simple exchange interaction form $f_{ex} = \exp(-\alpha r_{ij})$.

In this case, only 2 parameters need to be chosen, A and α. As a result of this, the effect of the magnetic modifications to the Lennard-Jones potential are to raise and/or lower the energy of aligned and anti-aligned spins, depending on the sign of A. If A is positive, pairs of aligned spins will act to lower the energy and anti-aligned spins will raise it. In this case, ferromagnetic magnetic structures are expected to have the lowest energy, since aligned spins would act to decrease the total energy of the system. If A is negative, pairs of anti-aligned spins will lower the energy and pairs of aligned spins will raise it. In this case, antiferromagnetic structures are expected to have the lowest energy, since aligned spins would act to increase the total energy of the system.

In order to get physically sensible values for A and α, they can be parameterised in a number of ways. The way which has been chosen here is to parameterise the values to the equilibrium distance r_{min} of an Fe dimer, along with the difference in energy at r_{min} and $2r_{min}$ between aligned and anti-aligned dimers, as demonstrated in Figure 7. These values have been found by performing a DFT calculation using the CASTEP code and fitting the results to the V^{LJ+S} potential form (Equation (6)).

3.1.1. Algorithm Performance

The magnetic GA was run on two different six-atom systems of the LJ+S potential. The first used the parameters listed in Table 1 and the other had the same parameters but with the sign of A inverted. This provides both a ferromagnetic and antiferromagnetic system on which to test the algorithm. The GA was allowed to optimise the atomic positions, spins and lattice parameters. The GA was run with population size of 24 members per generation, and ran for a maximum of 50 generations for the FM case and 80 generations for the AFM case.

Table 1. Table listing the parameters for the ferromagnetic (FM) parameterisation of the LJ + S potential.

Parameter	Value
$\varepsilon^{\text{LJ}+\text{S}}$	1.75 eV
$\sigma^{\text{LJ}+\text{S}}$	1.87 Å
A	−2.82 eV
α	0.83 Å^{-1}

Each member was locally optimised using the TPSD algorithm for 50 iterations of the local optimisation to get into the quadratic region, then further converged with the BFGS algorithm. The structures were converged to a tolerance of 1 meV. Since the LJ+S system spins could not be locally optimised by CASTEP, the GA spin mutation operation was modified to be a normally distributed rotation around the surface of the unit sphere, centring on the spin's previous position.

A mutation amplitude of 2 Å was used for the ions, at a rate of 0.03 mutations per atom. This was chosen to allow on average one atom every member to mutate, but significantly enough to hop lattice sites. The spin mutation rate was 0.06 per atom, since the mutations in the spin would be less dramatic.

Figure 10 shows the enthalpies of each structure as a function of the sum of the z component of the spins for both the AFM and FM parameterisations. It can be seen that, as the calculation progresses, both the enthalpies and spins of each system converge to their stable states, i.e., $\sum S_z = 0\,\hbar/2$ for the AFM case and $\sum S_z = 6\,\hbar/2$ for the FM case. Since the magnitude of the spins on each atom is kept constant at $|\vec{S}_i| = 1\hbar/2$, these states correspond to fully AFM and fully FM structures respectively.

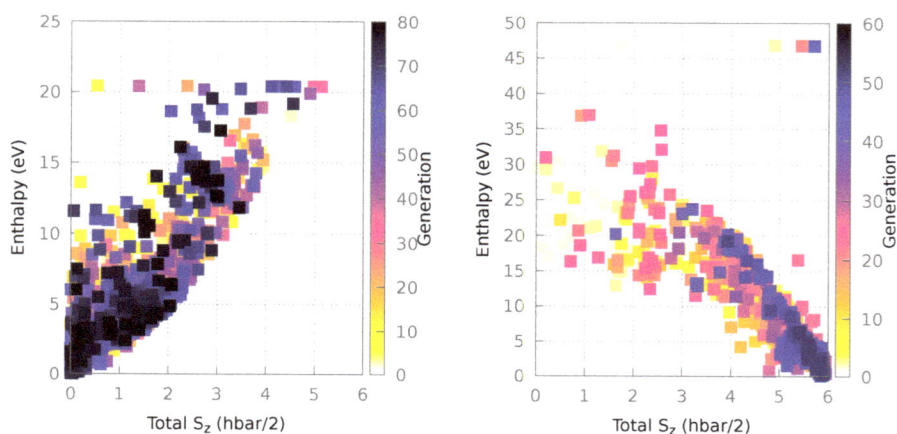

Figure 10. Convergence of the enthalpy and spin ($\sum S_z$) for each structure found by the GA relative to the ground state for the two parameterisations of the LJ + S potential: AFM (**left**) and FM (**right**). The color represents the generation in which that structure was found.

3.1.2. Final Structures

The final lowest enthalpy structures are shown in Figure 11. Common neighbour analysis by the software visualisation package Ovito [37] reveals the ground state structure for the AFM parameterisation to be face-centred cubic (FCC) and the ground state structure of the FM system to be body-centred cubic (BCC). It can be seen that as well as finding the crystal structures, the GA was able to correctly align and anti-align the spins for the FM/AFM structures respectively. This illustrates the power of our new magnetic GA to perform global optimisation in both spin and crystal structure spaces simultaneously.

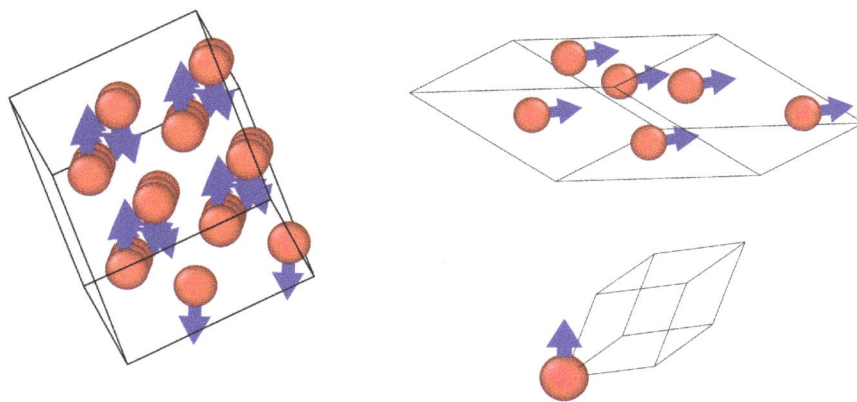

Figure 11. Lowest enthalpy structures for the AFM (**left**) and FM (**top right** and **bottom right**) parameterisations of the LJ + S model. For the FM parameterisation, both the six-atom unit cell (**top right**) and primitive cell (**bottom right**) are shown. For the AFM parameterisation, a $2 \times 2 \times 1$ supercell is shown to emphasise the BCC structure.

3.2. Heusler/Ge Interface

For the CFAS/Ge interface structure search, the calculation was performed on the 12-atom unit cell containing four germanium, four cobalt, two iron, one aluminium and one silicon atom. The GA was allowed to fully optimise the atomic positions. The lattice parameters were constrained to that of the full Heusler alloy, $a = b = c = 5.676$ Å, since there is very little lattice mismatch to the germanium lattice and it is not expected that this would change much over such a short interface region. The GA was run with a population size of 20 members per generation for a maximum of 100 generations.

An ionic position mutation amplitude of 3 Å was used with a rate of 0.03 mutations per atom. This was chosen to allow on average one atom every two members to mutate, but significantly enough to hop lattice sites. A slightly higher ionic permutation rate of 0.04 was chosen since, for the bulk Heusler, atomic swaps on the lattice sites can be comparable in energy to the ground state. If a similar crystal structure is found, it is likely that exploring this kind of disorder would be a good idea. The spin mutation rate was 0.06 per atom. This was chosen to be higher than the atomic mutation rate because not all atoms would be expected to magnetise, meaning that some mutations would get negated by the local optimisation.

Each member was optimised using the BFGS algorithm to a tolerance of 1 meV. The energies were computed using CASTEP, with a cutoff energy of 650 eV and k-point spacing of 0.04 Å^{-1}. The PBE functional [22] was used in conjunction with Hubbard U values of 2.1 eV on the *d*-orbitals of iron and cobalt, in keeping with previous CASTEP studies of the related Heusler alloys [35].

3.2.1. Algorithm Performance

Figure 12 shows the range of enthalpies for CFAS/Ge structures found at each generation of the GA. It can be seen that, on average, the enthalpy of structures in decreases significantly over the first 10 generations, and remains below that of the random structures searched in generation 0 for the remainder of the calculation. This implies that the GA successfully explores more favourable regions of the configuration space. In addition, it can be seen that every generation retains a significant distribution of structures over a range of about 2.5–3 eV. This suggests that the population is remaining significantly diverse and not converging to a single low enthalpy solution. Finally, the inset to Figure 12 shows how the lowest enthalpy value found changes during the calculation. The overall lowest enthalpy structure of the run was found in generation 72. The calculation proceeded for a further 28 generations without finding a lower enthalpy structure and hence this can be taken as a reasonable candidate for the global lowest enthalpy structure.

Figure 13 shows the distribution of magnetic structures found, plotted against their energy. Two clusters of structures are found in the energy range which would be thermally accessible during growth. The first of these are ferromagnetic, with a total spin of about 8 $\hbar/2$ and a modulus spin of about 9.9 $\hbar/2$. The second cluster is antiferromagnetic, with a total spin of about 0 $\hbar/2$ and a modulus spin of about 7.5 $\hbar/2$.

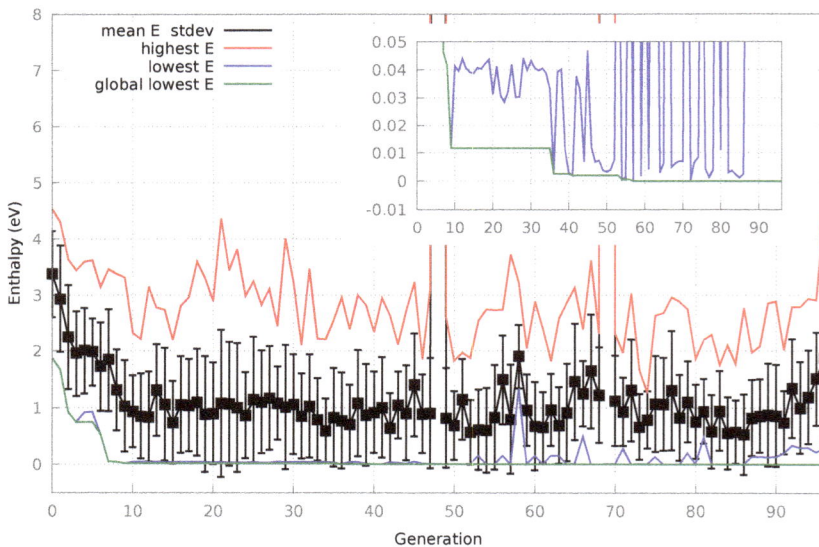

Figure 12. Total enthalpy per 12-atom cell of CFAS/Ge structure relative to the overall lowest enthalpy structure found. The lines represent the highest, lowest and mean enthalpies, along with the overall best structure found so far for each generation. The inset shows the best structures found in the 0–0.05 eV range.

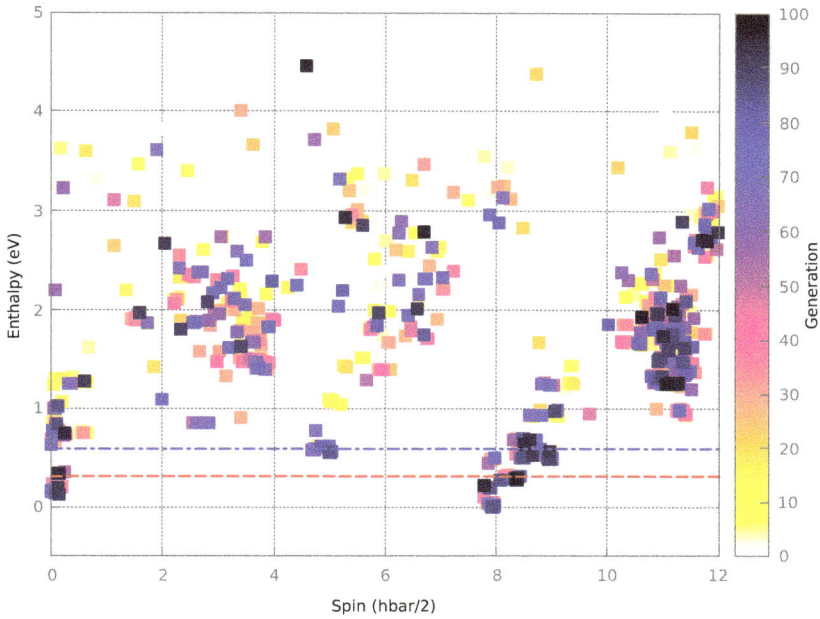

Figure 13. Total spin of the $Co_2FeAl_{0.5}Si_{0.5}$ (CFAS)/Ge GA structures, plotted against enthalpy of the 12 atom cell, relative to the overall lowest enthalpy structure found. Thermally accessible energies are marked for 575 K (blue dots) and 300 K (red dashes). The color of the points represents the generation in which the structure was found.

Figure 14 shows these regions in more detail. It can be seen that each cluster contains a number of structures. These are atomic swaps of the lowest enthalpy structure in each magnetic state. Candidate structures are labelled A1–A3 for the AFM structures and F1–F4 for the FM structures.

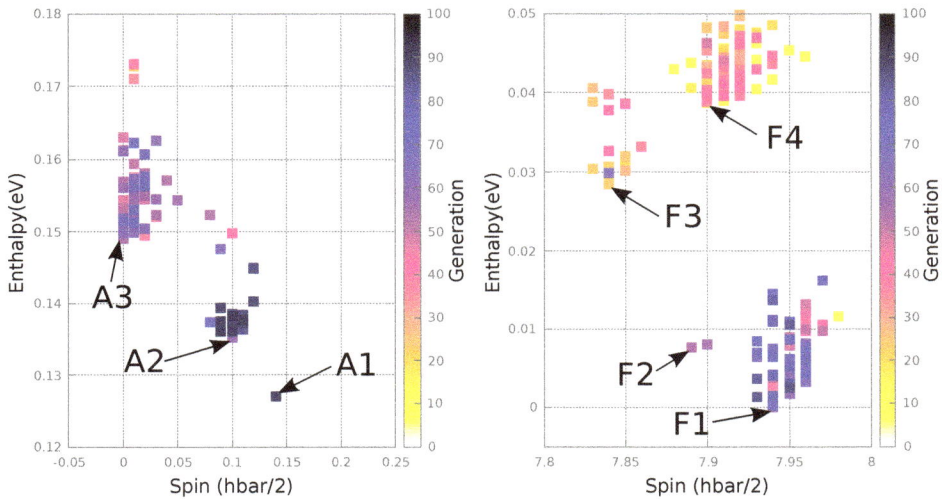

Figure 14. Clustering of CFAS/Ge structures around the AFM (**left**) and FM (**right**) configurations of the interface phase. In each case, a number of distinct structures have been identified (A1–3, F1–4).

3.2.2. Resultant Structures

The lowest enthalpy structure F1, shown in Figure 15 turns out to be a half Heusler alloy, where the X_1 site contains cobalt, the X_2 site contains a vacancy, the Y site contains the iron, silicon and aluminium, and the Z site contains germanium. The F2 structure has the same crystal structure as F1 but with a slightly different spin structure. The F3 and F4 structures have the same crystal structure except one germanium atom is swapped with the silicon atom from the Y site. The results are summarised in Table 2. We also performed a Hirshfeld charge and spin analysis, which showed that in each of these structures the spin is localised on the iron and cobalt, with a spin of around $3 \, \hbar/2$ on the iron atoms and $0.5 \, \hbar/2$ on the cobalt atoms.

Table 2. Table showing the total enthalpies and spins of the best structures, together with the lattice permutations in relation to F1.

Structure	Enthalpy (eV)	Total Spin ($\hbar/2$)	Total |Spin| ($\hbar/2$)	Disorder
F1	0.0	7.94	9.91	None
F2	0.01	7.89	9.89	None
F3	0.03	7.84	9.92	Ge $\leftrightarrow$ Si
F4	0.04	7.89	9.93	Ge $\leftrightarrow$ Si
A1	0.13	0.13	7.50	2Ge $\leftrightarrow$ Si,Fe
A2	0.14	0.10	7.45	Ge $\leftrightarrow$ Fe
A3	0.15	0.00	7.62	None

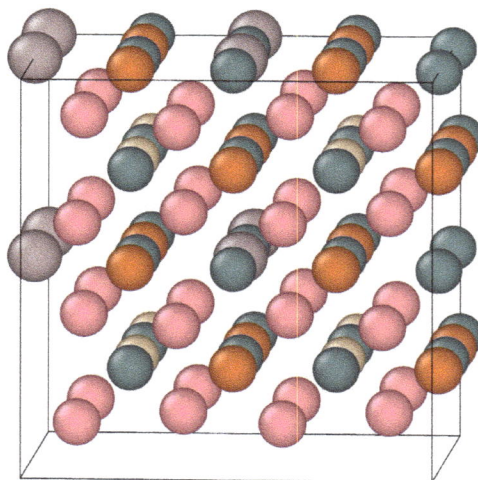

Figure 15. The $2 \times 2 \times 2$ unit cells of the lowest enthalpy structure found by the GA, showing the positions of the cobalt (pink), aluminium (grey), germanium (turquoise), silicon (gold) and iron (orange) atoms.

Unlike the full CFAS structure, none of these structures are half-metallic, with a significant number of states around the Fermi energy. It does not appear that the atomic disorder separating F1–4 has a significant effect on the electronic structure of the structures. Figure 16 shows the density of states for bulk CFAS, along with the lowest enthalpy FM and AFM structures found by the GA. It can be seen that, unlike the bulk CFAS structure, these structures are not half metallic, as there is no band gap in the minority spin channel. This provides fundamental insight into how this interface structure might degrade device performance.

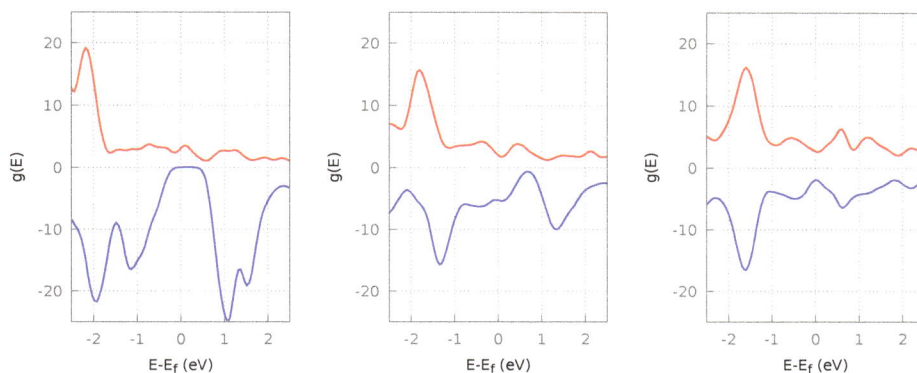

Figure 16. Density of states (DoS) for bulk CFAS (**left**) and two candidate half-Heuslers: F1 (**middle**) and A1 (**right**). The red and blue lines show the DoS for the majority and minority spins respectively relative to the Fermi energy. These are characteristic of the FM and AFM structures discussed in this paper. It can be seen that, unlike bulk CFAS, neither of the half-Heusler structures have a band gap in the minority spin channel.

This illustrates the power of our new magnetic GA in an ab initio context, to fully optimise spin and structure simultaneously, even in the absence of any experimental information on a candidate structure.

4. Conclusions

We have presented an enhanced GA for the structure prediction of magnetic materials, such that the magnetic and crystal structure may be predicted simultaneously. We have introduced new operations for atomic spin such as crossover, mutation and permutation operations. In addition, the idea of structural fingerprinting using the crystallographic structure factor was extended, based on ideas from magnetic neutron scattering theory, and it was demonstrated that the new fingerprint could identify both similar and distinct crystal and magnetic structures.

To test the magnetic GA, we introduced a novel pair potential V^{LJ+S} which included magnetic-like effects. While this potential does not accurately model any particular real-life material, it provided a computationally efficient way of exploring magnetic materials without the cost of fully quantum calculations. Two parameterisations were given for this potential. The first was parameterised by DFT simulations of Fe dimers in both aligned and anti-aligned configurations. Since the Fe dimers preferred to be aligned, it was expected that this parameterisation would yield a FM structure. This was observed as the lowest energy structure found by the GA was FM FCC, and there was a clear trend in the energies of all the structures searched towards FM structures. The second parameterisation was the same as the first, except for a reversed sign on the exchange-like interaction. This was expected to result in AFM structures since pairs of aligned spins would raise the energy of the system. Indeed, the lowest energy structure discovered by the GA was an AFM BCC structure. In addition, there was a clear trend amongst the other structures towards AFM alignment.

Finally, as an example of studying novel magnetic structures of significant experimental interest, the interface between the Heusler alloy CFAS and n-doped germanium was investigated. Experimentally, the two materials showed significant mixing and, when annealed, a new phase formed at the interface as shown by EDS measurements. The structure of this phase was investigated with the new magnetic GA, and a half-Heusler structure was predicted.

To conclude, we have presented a new GA for structure prediction capable of optimising both magnetic and crystal structures simultaneously.

Author Contributions: The different contributions of the authors to this study were as follows: conceptualisation, M.I.J.P.; methodology, E.J.H., P.J.H., M.I.J.P.; software, E.J.H., M.I.J.P.; manuscript preparation, E.J.H., P.J.H., M.I.J.P.

Funding: E.J.H. was financially supported by EPSRC under the DTA scheme. Computing resources on ARCHER were provided by UKCP under EPSRC grant EP/P022561/1. P.J.H. was funded by EPSRC grant EP/R025770/1.

Acknowledgments: The authors wish to acknowledge Balati Kuerbanjiang and Vlado Lazarov for providing the EDS experimental results and for discussion on the CFAS study.

Conflicts of Interest: The authors declare no conflict of interest. The funders had no role in the design of the study; in the collection, analyses, or interpretation of data; in the writing of the manuscript, or in the decision to publish the results.

Abbreviations

The following abbreviations are used in this manuscript:

CFAS	$Co_2FeAlSi$ (cobalt iron aluminium silicide; a half-metallic Heusler alloy)
DFT	Density Functional Theory
GA	Genetic Algorithm
LJ	Lennard-Jones
LJ+S	Lennard-Jones with spin
EDS	Energy dispersive X-ray spectroscopy
DoS	Density of States

References

1. Comstock, R.L. Review modern magnetic materials in data storage. *J. Mater. Sci. Mater. Electron.* **2002**, *13*, 509–523. [CrossRef]
2. Weller, D.; Parker, G.; Mosendz, O.; Lyberatos, A.; Mitin, D.; Safonova, N.Y.; Albrecht, M. Review article: FePt heat assisted magnetic recording media. *J. Vac. Sci. Technol. B* **2016**, *34*, 060801. [CrossRef]
3. Balke, B.; Wurmehl, S.; Fecher, G.H.; Felser, C.; Kübler, J. Rational design of new materials for spintronics: Co_2FeZ (Z = Al, Ga, Si, Ge). *Sci. Technol. Adv. Mater.* **2008**, *9*, 014102. [CrossRef] [PubMed]
4. Li, X.; Yang, J. First-principles design of spintronics materials. *Natl. Sci. Rev.* **2016**, *3*, 365–381. [CrossRef]
5. Coey, J. Rare-earth magnets. *Endeavour* **1995**, *19*, 146–151. [CrossRef]
6. Alonso, E.; Sherman, A.M.; Wallington, T.J.; Everson, M.P.; Field, F.R.; Roth, R.; Kirchain, R.E. Evaluating rare earth element availability: A case with revolutionary demand from clean technologies. *Environ. Sci. Technol.* **2012**, *46*, 3406–3414. [CrossRef] [PubMed]
7. Wales, D.J.; Scheraga, H.A. Global optimization of clusters, crystals, and biomolecules. *Science* **1999**, *285*, 1368–1372. [CrossRef] [PubMed]
8. Amsler, M.; Goedecker, S. Crystal structure prediction using the minima hopping method. *J. Chem. Phys.* **2010**, *133*, 224104. [CrossRef]
9. Panosetti, C.; Krautgasser, K.; Palagin, D.; Reuter, K.; Maurer, R.J. Global materials structure search with chemically motivated coordinates. *Nano Lett.* **2015**, *15*, 8044–8048. [CrossRef]
10. Pickard, C.J.; Needs, R.J. Ab initio random structure searching. *J. Phys. Condens. Matter* **2011**, *23*, 053201. [CrossRef]
11. Woodley, S.M.; Catlow, R. Crystal structure prediction from first principles. *Nat. Mater.* **2008**, *7*, 937–946. [CrossRef] [PubMed]
12. Wang, Y.; Lv, J.; Zhu, L.; Ma, Y. Crystal structure prediction via particle-swarm optimization. *Phys. Rev. B* **2010**, *82*, 094116. [CrossRef]
13. Woodley, S.M.; Battle, P.D.; Gale, J.D.; Richard, A.; Catlow, C. The prediction of inorganic crystal structures using a genetic algorithm and energy minimisation. *Phys. Chem. Chem. Phys.* **1999**, *1*, 2535–2542. [CrossRef]
14. Abraham, N.L.; Probert, M.I.J. A periodic genetic algorithm with real-space representation for crystal structure and polymorph prediction. *Phys. Rev. B* **2006**, *73*, 224104. [CrossRef]
15. Glass, C.W.; Oganov, A.R.; Hansen, N. USPEX—Evolutionary crystal structure prediction. *Comput. Phys. Commun.* **2006**, *175*, 713 – 720. [CrossRef]
16. Lonie, D.C.; Zurek, E. XtalOpt: An open-source evolutionary algorithm for crystal structure prediction. *Comput. Phys. Commun.* **2011**, *182*, 372–387. [CrossRef]

17. Lennard-Jones, J.E. Cohesion. *Proc. Phys. Soc.* **1931**, *43*, 461. [CrossRef]
18. Hohenberg, P.; Kohn, W. Inhomogeneous electron gas. *Phys. Rev.* **1964**, *136*, B864–B871. [CrossRef]
19. Kohn, W.; Sham, L.J. Self-Consistent equations including exchange and correlation effects. *Phys. Rev.* **1965**, *140*, A1133–A1138. [CrossRef]
20. Payne, M.C.; Teter, M.P.; Allan, D.C.; Arias, T.A.; Joannopoulos, J.D. Iterative minimization techniques for ab initio total-energy calculations: Molecular dynamics and conjugate gradients. *Rev. Mod. Phys.* **1992**, *64*, 1045–1097. [CrossRef]
21. Woods, N.; Payne, M.; Hasnip, P. Computing the self-consistent field in Kohn-Sham density functional theory. *J. Phys. Condens. Matter* **2019**, *31*, 453001. [CrossRef] [PubMed]
22. Perdew, J.P.; Burke, K.; Ernzerhof, M. Generalized gradient approximation made simple. *Phys. Rev. Lett.* **1996**, *77*, 3865–3868. [CrossRef] [PubMed]
23. Jones, R.O.; Gunnarsson, O. The density functional formalism, its applications and prospects. *Rev. Mod. Phys.* **1989**, *61*, 689–746. [CrossRef]
24. Hasnip, P.J.; Refson, K.; Probert, M.I.J.; Yates, J.R.; Clark, S.J.; Pickard, C.J. Density functional theory in the solid state. *Philos. Trans. R. Soc. A Math. Phys. Eng. Sci.* **2014**, *372*. [CrossRef] [PubMed]
25. Abraham, N.; Probert, M. Improved real-space genetic algorithm for crystal structure and polymorph prediction. *Phys. Rev. B* **2008**, *77*, 134117. [CrossRef]
26. Abraham, N.L.; Probert, M.I.J. Erratum: Improved real-space genetic algorithm for crystal structure and polymorph prediction [Phys. Rev. B 77, 134117 (2008)]. *Phys. Rev. B* **2016**, *94*, 059904. [CrossRef]
27. Kübler, J. Magnetic moments of ferromagnetic and antiferromagnetic bcc and fcc iron. *Phys. Lett. A* **1981**, *81*, 81–83. [CrossRef]
28. Hirshfeld, F.L. Bonded-atom fragments for describing molecular charge densities. *Theor. Chim. Acta* **1977**, *44*, 129–138. [CrossRef]
29. Clark, S.J.; Segall, M.D.; Pickard, C.J.; Hasnip, P.J.; Probert, M.I.; Refson, K.; Payne, M.C. First principles methods using CASTEP. *Z. Für Krist.-Cryst. Mater.* **2005**, *220*, 567–570. [CrossRef]
30. Bacon, G. *Neutron Diffraction Monographs on the Physics and Chemistry of Materials*; Oxford University Press: New York, NY, USA, 1975.
31. Harrison, R.J.; Krasko, G.L. Magnetic-state-dependent interatomic potential for iron (abstract). *J. Appl. Phys.* **1990**, *67*, 4585. [CrossRef]
32. Dudarev, S.L.; Derlet, P.M. A 'magnetic' interatomic potential for molecular dynamics simulations. *J. Phys. Condens. Matter* **2005**, *17*, 7097. [CrossRef]
33. Ackland, G.J. Two-band second moment model for transition metals and alloys. *J. Nucl. Mater.* **2006**, *351*, 20–27. [CrossRef]
34. Dragoni, D.; Daff, T.D.; Csányi, G.; Marzari, N. Achieving DFT accuracy with a machine-learning interatomic potential: Thermomechanics and defects in bcc ferromagnetic iron. *Phys. Rev. Mater.* **2018**, *2*, 013808. [CrossRef]
35. Nedelkoski, Z.; Hasnip, P.J.; Sanchez, A.M.; Kuerbanjiang, B.; Higgins, E.; Oogane, M.; Hirohata, A.; Bell, G.R.; Lazarov, V.K. The effect of atomic structure on interface spin-polarization of half-metallic spin valves: Co_2MnSi/Ag epitaxial interfaces. *Appl. Phys. Lett.* **2015**, *107*, 212404. [CrossRef]
36. Kuerbanjiang, B.; Fujita, Y.; Yamada, M.; Yamada, S.; Sanchez, A.M.; Hasnip, P.J.; Ghasemi, A.; Kepaptsoglou, D.; Bell, G.; Sawano, K.; et al. Correlation between spin transport signal and Heusler/semiconductor interface quality in lateral spin-valve devices. *Phys. Rev. B* **2018**, *98*, 115304. [CrossRef]
37. Stukowski, A. Visualization and analysis of atomistic simulation data with OVITO—The open visualization tool. *Model. Simul. Mater. Sci. Eng.* **2009**, *18*, 015012. [CrossRef]

crystals

MDPI

Article

Quest for Compounds at the Verge of Charge Transfer Instabilities: The Case of Silver(II) Chloride [†]

Mariana Derzsi [1,2,*], Adam Grzelak [1], Paweł Kondratiuk [1,‡], Kamil Tokár [2,3] and Wojciech Grochala [1,*]

[1] Center of New Technologies, University of Warsaw, Zwirki i Wigury 93, 02089 Warsaw, Poland
[2] Advanced Technologies Research Institute, Faculty of Materials Science and Technology in Trnava, Slovak University of Technology in Bratislava, 917 24 Trnava, Slovakia
[3] Institute of Physics, Slovak Academy of Sciences, 845 11 Bratislava, Slovakia
[*] Correspondence: mariana.derzsi@gmail.com (M.D.); w.grochala@cent.uw.edu.pl (W.G.)
[†] This work is dedicated to the memory of Kazimierz Fajans (1887–1975).
[‡] Current Address: Institute of Theoretical Physics, Faculty of Physics, University of Warsaw, 02089 Warsaw, Poland.

Received: 5 July 2019; Accepted: 9 August 2019; Published: 15 August 2019

Abstract: Electron-transfer processes constitute one important limiting factor governing stability of solids. One classical case is that of CuI_2, which has never been prepared at ambient pressure conditions due to feasibility of charge transfer between metal and nonmetal ($CuI_2 \rightarrow CuI + \frac{1}{2} I_2$). Sometimes, redox instabilities involve two metal centers, e.g., AgO is not an oxide of divalent silver but rather silver(I) dioxoargentate(III), $Ag(I)[Ag(III)O_2]$. Here, we look at the particularly interesting case of a hypothetical $AgCl_2$ where both types of redox instabilities operate simultaneously. Since standard redox potential of the Ag(II)/Ag(I) redox pair reaches some 2 V versus Normal Hydrogen Electrode (NHE), it might be expected that Ag(II) would oxidize Cl^- anion with great ease (standard redox potential of the $\frac{1}{2}$ Cl_2/Cl^- pair is + 1.36 V versus Normal Hydrogen Electrode). However, ionic $Ag(II)Cl_2$ benefits from long-distance electrostatic stabilization to a much larger degree than $Ag(I)Cl + \frac{1}{2} Cl_2$, which affects relative stability. Moreover, Ag(II) may disproportionate in its chloride, just like it does in an oxide; this is what $AuCl_2$ does, its formula corresponding in fact to $Au(I)[Au(III)Cl_4]$. Formation of polychloride substructure, as for organic derivatives of Cl_3^- anion, is yet another possibility. All that creates a very complicated potential energy surface with a few chemically distinct minima i.e., diverse polymorphic forms present. Here, results of our theoretical study for $AgCl_2$ will be presented including outcome of evolutionary algorithm structure prediction method, and the chemical identity of the most stable form will be uncovered together with its presumed magnetic properties. Contrary to previous rough estimates suggesting substantial instability of $AgCl_2$, we find that $AgCl_2$ is only slightly metastable (by 52 meV per formula unit) with respect to the known AgCl and $\frac{1}{2}$ Cl_2, stable with respect to elements, and simultaneously dynamically (i.e., phonon) stable. Thus, our results point out to conceivable existence of $AgCl_2$ which should be targeted via non-equilibrium approaches.

Keywords: silver; chlorine; learning algorithms; crystal structure; magnetic properties

1. Introduction

Electron-transfer processes constitute one important limiting factor governing stability of solids. One classic case is that of CuI_2, which has never been prepared at ambient pressure conditions due to feasibility of charge transfer between metal and nonmetal. The energy of ligand-to-metal-charge transfer (LMCT) is negative for CuI_2 which results in instability and phase separation, according to the Equation (1):

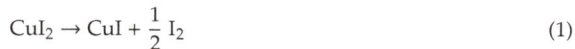

$$CuI_2 \rightarrow CuI + \frac{1}{2} I_2 \tag{1}$$

While the process of oxidation of iodide anions by Cu(II) is well-known to every chemistry freshman, it remains somewhat difficult to explain to a comprehensive school pupil, based on the values of standard redox potentials, E^0, for the relevant species in aqueous solutions. The reason for that is that the E^0 value for the Cu(II)/Cu(I) redox pair is + 0.16 V versus NHE (Normal Hydrogen Electrode), while that for the $I_2/2\,I^-$ is + 0.54 V, i.e., I_2 is formally a slightly better stronger oxidizer than Cu(II). The detailed explanation necessitates departure from the aqueous conditions (note, E^0 values are the feature of solvated species in aqueous solutions). Since Cu(II) in aqueous solutions is very strongly solvated by water molecules acting as a Lewis base ($H_2O \rightarrow Cu(II)$), and this effect surpassed the one of coordination of water molecules to Cu(I), the oxidizing properties of naked Cu(II) are certainly stronger than those of solvated ions. Simultaneously, since I^- anion is coordinated by water molecules acting like Lewis acid ($OH_2 \ldots I^-$) (and coordination to anion is stronger than to neutral I_2 molecules) the E^0 value for the $I_2/2\,I^-$ redox pair is actually larger than the one for unsolvated species. This helps to explain why the balance of a redox reaction for the phases in the solid state, i.e., lacking any solvent, is different from the one which might be guessed based on the plain E^0 values.

Attempts to more quantitatively explain the lack of stability of CuI_2 at ambient (p,T) conditions involve discussions of ionization potential of monovalent metal, ionic polarizabilities, as well as lattice energies of relevant solids. Over 60 years ago, following early considerations by Fajans [1], Morris estimated the standard molar free energy of formation of CuI_2 [2]. The obtained value was slightly positive, some 1 kcal/mole [2], see also [3]. Since the respective value for CuI is large and negative (circa −16.6 kcal/mole), it is natural that formation of CuI_2 is strongly disfavored. This, of course, may change under high pressure conditions, which—due to beneficial pV factor allowing to crowd cations and anions together in the lattice—often favor formation of species at high oxidation states.

Sometimes redox instabilities involve two metal centers rather than metal and nonmetal, e.g., AgO is not an oxide of divalent silver but rather silver(I) dioxoargentate(III), $Ag(I)[Ag(III)O_2]$ [4,5]:

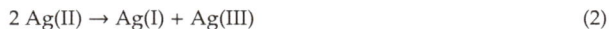

$$2\,Ag(II) \rightarrow Ag(I) + Ag(III) \tag{2}$$

The analogous behavior has been theoretically predicted for AuO, as well [6]. It is important to notice that the disproportionation reactions of this type are always energy uphill in the gas phase, and they are relatively rare for extended solids. In the case of silver, the energy of reaction proceeding according to Equation (2) is positive and very large, some 13.3 eV, as may be estimated from relevant ionization potentials (this value corresponds to the Mott-Hubbard U energy in the gas phase). The fact that the process takes place in AgO according to Equation (2) is given by several important factors, such as departure from ionic formulation, and the fact that U is strongly screened in solids. The rule of a thumb is that disproportionation processes are facile (i) in a Lewis-basic environment, (ii) especially when there is strong mixing of metal and nonmetal states i.e., pronounced covalence of chemical bonding, (iii) when the pV factor at elevated pressure, which prefers packing of unequal spheres, dominates the energetic terms, and (iv) at low temperatures [7]. AgO is in fact a nice exemplification of condition (i), since it is disproportionated, while its more Lewis acidic derivative, $AgSO_4$, is not [8]. Another good example is that of $Au(II)(SbF_6)_2$, which is, comproportionated, a genuine Au(II) compound [9], while its parent basic fluoride, AuF_2, has never been prepared in the solid phase, as it is subject to phase separation via disproportionation to Au and AuF_3. Moreover, AgO exemplifies condition (ii), since the chemical bonding between Ag and O is remarkably covalent in this compound, which leads to a phonon-driven disproportionation [10]. Finally, AgO, also exemplifies condition (iii), since it remains disproportionated to a pressure of at least 1 mln atm [11].

Here, we look at the particularly interesting case of an elusive $AgCl_2$ where, as we will see, both types of redox instabilities may happen. Since the standard redox potential of the Ag(II)/Ag(I) redox pair reaches some +2.0 V versus NHE and the respective value for the $\frac{1}{2}$ Cl_2/Cl^- pair is +1.36 V, it might naturally be expected that Ag(II) would oxidize Cl^- anion with great ease (the above-mentioned arguments valid for CuI_2 are also valid for $AgCl_2$). However, ionic $Ag(II)Cl_2$ benefits from long-distance electrostatic stabilization to a much larger degree than $Ag(I)Cl + \frac{1}{2}$ Cl_2, which should affect relative stability. Moreover, since the Ag(II)–Cl bonding is naturally expected to be quite covalent, similarly to the Ag(II)–O one in AgO (Cl and O have nearly identical electronegativities), Ag(II) might disproportionate in its chloride, just like it does in oxide. This is in fact what related $AuCl_2$ does, its true formula corresponding in fact to $Au(I)[Au(III)Cl_4]$ [12]. Formation of polychloride substructure, as for organic derivatives of Cl_3^- anion [13,14], or a mixed chloride-polychloride ($Ag(I)_2(Cl)(Cl_3)$), is yet another possibility. Last but not the least, $AgCl_2$ featuring an unpaired electron at the transition metal center may choose to form exotic Ag–Ag bond, as is observed for $AuSO_4$ [15]. All that creates a complicated potential energy surface with a few chemically distinct minima, i.e., diverse polymorphic forms present, and this renders theoretical predictions troublesome.

The major aim of this work is to theoretically predict crystal structure, stability, and presumed electronic and magnetic properties of the most stable form of $AgCl_2$. We would like also to computationally verify early predictions by Morris who estimated the free enthalpy of reaction:

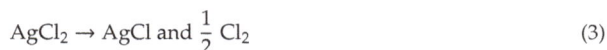

$$AgCl_2 \rightarrow AgCl \text{ and } \frac{1}{2} Cl_2 \qquad (3)$$

to be strongly negative, some −96.4 kJ/mole [2]. Finally, we will briefly discuss the anticipated impact of elevated pressure on the course of the reaction described by Equation (3), as well as magnetic properties of the most stable phases found.

2. Methodology

This study has begun in resemblance with our previous theoretical search for $AgSO_4$, where we have employed the method of following imaginary phonon modes to reach dynamically stable structural models [16,17]. Using this method, we have optimized the hypothetical $AgCl_2$ crystal in all known crystal structure polytypes taken by metal dihalides, MX_2 (X = F, Cl, Br, I) (i.e., over 40 structure types) using the plane-wave code VASP (Vienna Ab-initio Simulation Package) [18–22], and subsequently calculated phonon dispersion curves for each model using the program PHONON [23]. In cases where at least one imaginary phonon was detected (signaling dynamic instability), we followed the normal coordinate of this mode to reach another, lower symmetry and lower energy structure [17,24], and we reexamined phonons after optimization. Due to complexity and CPU burden of the task this preliminary quest was conducted with LDA functional and lower plane-wave cut-off equal to 400 eV. In the second approach, the evolutionary algorithm approach was applied using XtalOpt [25–28] in combination with VASP and using GGA (general gradient approximation) with Perdew-Burke-Ernzerhof functional adapted for solids (PBEsol) and plane-wave cutoffs of 520 eV. Using the evolutionary algorithms, we have considered unit cells containing 2, 4, and 8 formula units and generated a pool of 2183 structures. The lowest energy structures originating from both methods were ultimately recalculated with spin-polarization, on-site Coulomb interactions and van der Waals corrections, as outlined below. Magnetic models were constructed for polymorphic forms containing genuine Ag(II) paramagnetic centers and the lowest energy spin arrangements were found using the rotationally invariant density functional theory DFT + U method introduced by Liechtenstein et al. where the values of both Hubbard U and the Hund J parameter are set explicitly [29]. The U and J parameters were set only for d orbitals of the Ag atoms and the values of 5 eV and 1 eV were used, respectively [30]. The van der Waals interactions were accounted for using the DFT-D3 correction method of Grimme et al. with Becke-Jonson damping [31]. Additionally, the hybrid DFT calculations with HSE06 functional were used to calculate the mixed-valence $Ag^I Ag^{III} Cl_2$ solution that could not be properly stabilized at the DFT level.

Electronic density of states was calculated using the aforementioned DFT + U method with DFT-D3 van der Waals correction, with k-mesh of 0.025 Å^{-1} and 800 eV plane-wave cutoff. The disproportionated $AgCl_2$ and $AuCl_2$ structures were additionally pre-optimized with HSE06 functional using a coarser k-mesh of 0.05 Å^{-1}.

3. Results

3.1. Scrutiny of Dynamically Stable Polymorphic forms of $AgCl_2$ (Method of Following Imaginary Phonon Modes)

The crystal structures selected for preliminary study accounted for numerous polytypes known among the transition metal and alkali earth metal dihalides. Both more ionic as well as more covalent structural types were tested. AgF_2, $AuCl_2$, $CuCl_2$, PdF_2, $PtCl_2$, $PbCl_2$ (cottunite), α–PbO_2, and TiO_2 (rutile) were selected because of obvious structural analogies within Group 11 of the Periodic Table of Elements, or because they are often adopted by metal chlorides [24]. Among those, $KAuF_4$ and $AuCl_2$ types with K or Au atoms substituted by Ag ones, represent disproportionated Ag(I)/Ag(III) systems; others correspond to comproportionated ones. We have also employed a set of ionic halide structures, notably: CaF_2 (fluorite), $CaCl_2$ ($Pmn2_1$ and $Pnnm$ polytypes), $CdCl_2$, layered CdI_2, $MgCl2$ ($P4m2$, $Ama2$ and P–1 polymorphs), SrI_2, $YbCl_2$, as well as polymeric BeF_2, and three covalent structures: SiS_2, FeP_2, and XeF_2. Altogether, these representative prototypes show a rich variety of structural motifs and lattice dimensionalities. Using the method of following the imaginary modes of $AgCl_2$ in these types of structures we have obtained over 10 dynamically stable structures. Figure 1a–f illustrates the six main structural motives present in them. All remaining predicted polymorphs are simply various polytypes of these (differ in stacking of the main structural motives).

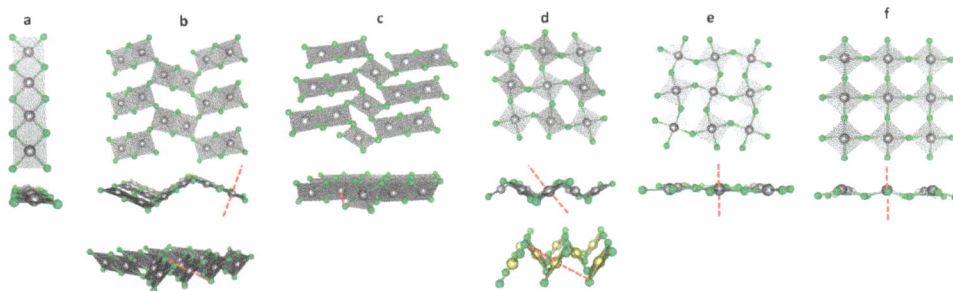

Figure 1. Illustration of main structural motifs that differ in connectivity of coordination polyhedra (the [$AgCl_4$] plaquettes) for the dynamically stable polymorphs of a hypothetical $AgCl_2$ compound predicted using the method of following imaginary phonon modes. Top and side view is shown. In the side views, the axial Ag-Cl contacts that complete the octahedral coordination of silver atoms are indicated by red dotted line. In panel **b**, two side views represent two distinct polymorphs, respectively. The lower side view in panel **d** (bottom), represents highly puckered layer that is observed in some transition metal dihalides, but not in $AgCl_2$ (see text for further explanation). Color code: Ag—big grey balls, Cl—small green balls, Pd—big yellow balls.

The main structural building block in all dynamically stable models is a [$AgCl_4$] plaquette. Here, silver is stabilized in a close to square-planar (or elongated octahedral) coordination by chloride anions, which is the most common coordination sphere of Ag^{II} cation among the known compounds [32]. Silver in the second oxidation state is seldom found in a linear (or a contracted octahedral) coordination, and other geometries are even more scarce. This behavior is nicely reflected by the results of our extensive structure screening. For example, the compressed octahedral coordination appeared in our search only once in a rutile type structure. Although predicted to be dynamically stable at DFT level, it was ruled out in the subsequent spin-polarized DFT + U calculations, where it converged to a square-planar coordination. The 2 + 4 coordination is indeed more common for fluorides in rutile structure (PdF$_2$ and NiF$_2$), but no such chloride is known. In one case, a butterfly penta-coordination (i.e., close to a tetragonal pyramide) was obtained, where the silver atoms are displaced out of the plain formed by the [AgCl$_4$] plaquettes (Figure 1f). Such geometry has been previously observed for Ag^{II} in two high-pressure polymorphs of AgF$_2$: a layered and a tubular one [33]. In both, the silver atoms depart from the center of the ideally flat square-planar [AgF$_4$] units to achieve batter packing while simultaneously preserving the local Jahn–Teller distortion. The AgCl$_2$ structure with the butterfly silver coordination is topologically equivalent to the layered HP1 polymorph of AgF$_2$. Indeed, it exhibits the lowest calculated volume among the predicted dynamically stable structures and thus it should be stabilized at high pressure (see section 3). Our scrutiny of AgCl$_2$ polytypes provides theoretical evidence that the butterfly coordination is indeed a natural response of octahedral Ag^{II} sites (4 + 2 coordination) to high pressures and it permits more effective packing of 4 + 1 + 1 distorted [Ag^{II}X$_6$] units.

The [Ag^{II}X$_4$] squares show three distinct connectivity patterns in the dynamically stable polymorphs. They are connected either by corners, edges or via a combination of the two, while the resulting lattice is one- or two-dimensional at most. This comes as no surprise since it is natural for the strongly Jahn-Teller active cation to exhibit reduced structural dimensionality in its compounds. Here, the edge sharing always results in one-dimensional chains that have a shape of infinite molecular ribbons (Figure 1a). On the other hand, corner and combined corner plus edge sharing leads always to layered structures (Figure 1b–f). No polymorphs with isolated [AgCl$_4$] units or three-dimensional connectivity were found. Additionally, it may be noticed that each chlorine atom is always shared between two Ag cations thus AgCl$_2$ strictly avoids chlorine terminals in its structures. This, too, is quite natural, since Ag^{II} is an electron deficient and Lewis acidic cation, which attempts to satisfy its need for electronic density by having at least four anions in its coordination sphere; at AgCl$_2$ stoichiometry this implies ligand sharing, i.e., [AgCl$_{4/2}$]. The charge depletion on Cl atoms affects the halogen-halogen interactions, as discussed in Supplementary Materials (S1).

Structures with infinite ribbons are characteristic of dihalides containing Jahn-Teller active cations and are also present in cuprates such as LiCu$_2$O$_2$ [34] and LiCuVO$_4$ [35]. In halides, the ribbons have neutral charge and these structures are held together by van der Waals interactions. In the cuprates, the ribbons are present as anionic species [CuO$_2$]$^{2-}$$_\infty$, whose charge is compensated by the presence of additional metal cations. Although observed in majority of halides containing Jahn-Teller ions including CuCl$_2$, CuBr$_2$, PdCl$_2$, PtCl$_2$, and CrCl$_2$, they have never been observed in compounds of silver. Importantly, for AgCl$_2$ this structure polytype has the lowest computed energy as will be discussed later in the text.

Three district structural patterns are observed among the layered polymorphs. In the first case, fragments of the ribbons may be distinguished that consist of two [AgCl$_4$] units sharing one edge. These [Ag$_2$Cl$_6$] dimers then interconnect into layers by sharing corners (Figure 1b). Another structural pattern is formed by alternation of the same dimmers with single squares (Figure 1c). The third type of layers is formed by squares sharing only corners (Figure 1d–f). The layered polymorphs containing the dimeric units are unique among the halides. They are closely related to orthorhombic ramsdellite [36] and monoclinic γ-MnO$_2$ polymorph form [37]. The ramsdellite structure consists of three-dimensional network of double chains of edge-sharing MnO$_6$ octahedra while in the γ-MnO$_2$ the double chains alternate with single chains of MnO$_6$ octahedra (Figure 2a,b). In the predicted AgCl$_2$ polymorphs with the dimeric [Ag$_2$Cl$_6$] units (Figure 1b), the three-dimensional network of the ramsdellite structure is reduced to two-dimensional one due to Jahn-Teller distortion of the octahedra that takes place in the direction parallel to the propagation of the chains. The same relation exists between the AgCl$_2$ polymorph formed by alternation of the [Ag$_2$Cl$_6$] dimers with [AgCl$_4$] squares (figure 1c) and the γ-MnO$_2$ structure. Although no such halides exist, the ramsdellite-related AgCl$_2$ structure has its zero-dimensional analogues in 4d and 5d transition metal pentachlorides such as MoCl$_5$, Ta$_2$Cl$_{10}$, NbCl$_5$, WCl$_5$. They consist of dimeric M$_2$Cl$_{10}$ units of edge-sharing MCl$_6$ octahedra aligned into infinite chains (Figure 2c). One can imagine obtaining the ramsdellite and the related layered $AgCl_2$ structure by virtual polymerization of the M_2Cl_{10} dimmers and subsequent Jahn-Teller distortion, respectively. While the layered structure containing the dimeric $[M_2Cl_6]$ units are to best of our knowledge unknown, the layers with corner-sharing of square planar $[MCl_4]$ units are well documented for transition metal halides including CuF_2, AgF_2, or $PdCl_2$; thus, it is quite natural to detect them for related $AgCl_2$.

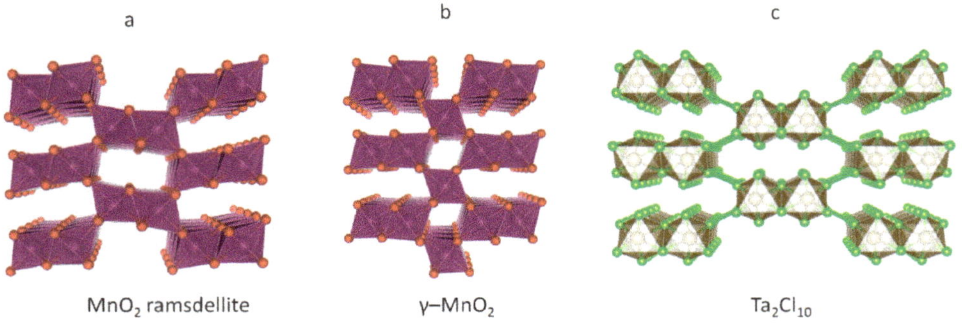

a

b

c

MnO$_2$ ramsdellite

γ–MnO$_2$

Ta$_2$Cl$_{10}$

Figure 2. Crystal structure of (**a**) ramsdellite, (**b**) γ-MnO_2 polymorph, and (**c**) that of molecular Ta_2Cl_{10} crystal. In case of the MnO_2 forms, the three-dimensional network of MnO_6 octahedra is highlighted. In case of the molecular Ta_2Cl_{10} crystal, selected van der Waals Cl . . . Cl contacts between the isolated edge-sharing MnO_6 octahedra are shown to highlight its relation to the ramsdellite structure.

All polymorphs predicted in this study are related to three archetypical structures, namely CdI_2, rutile, and fluorite structure. All ribbon polymorphs may be derived from the layered CdI_2 prototype, where each layer is formed by edge-sharing $[CdI_6]$ octahedra. Replacing the octahedral cadmium cation by a Jahn-Teller active one leads to elongation of octahedra and dissociation of the layers into infinite ribbons as illustrated in Figure 3c. In fact, the CdI_2-type layers and ribbons are the most common structural motifs among transition metal dihalides. Notably, all dichlorides of 3d elements crystallize in the CdI_2 polytypes, the only exception being those containing Jahn-Teller active ions, which in turn crystallize in ribbon structures.

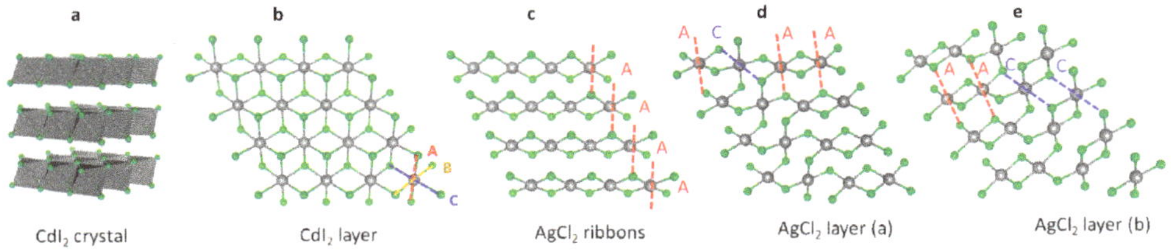

a

b

c

d

e

CdI$_2$ crystal

CdI$_2$ layer

AgCl$_2$ ribbons

AgCl$_2$ layer (a)

AgCl$_2$ layer (b)

Figure 3. Derivation of the predicted ribbon and layered $AgCl_2$ polymorphs by orbital ordering of the Jahn-Teller (JT) active dz^2 orbitals in the CdI_2-type structure. The original CdI_2 structure is shown in (**a**) and the single layer in (**b**). Orientation of the JT orbitals manifests itself by direction of octahedral elongation denoted by red dashed lines. Resulting orbital ordering patterns are also indicated for $AgCl_2$ ribbons—AAA as shown in (**c**), $AgCl_2$ layer (**a**)—AACAAC as shown in (**d**), and $AgCl_2$ layer (**b**)—AACC as shown in (**e**).

The layered polymorphs containing the $[Ag_2Cl_6]$ dimers can be also derived from the CdI_2 structure. As already emphasized, the ribbon structure may be obtained from the CdI_2 structure simply by elongation of the $[CdI_6]$ octahedra. This elongation is a consequence of Jahn-Teller stabilization (expansion) of the dz^2 orbitals, which may in principle be realized along any of the three main octahedral axes denoted by letters A, B and C in Figure 3b. This gives way to various possible orbital ordering patterns and thus various types of connectivity of the [$AgCl_4$] plaquettes. While the same orientation of the dz^2 orbitals along the A direction (ferrodistortive AAA orbital ordering pattern) leads to the ribbon polymorphs, alternating orientation of the orbitals (antiferrodistortive ordering patterns) results in layered polymorphs featuring corner-shared [Ag_2Cl_6] dimers. Here, the Jahn-Teller distortion takes place alternatively along the B and C direction while two such orbital ordering patterns are possible. The AACAAC orbital ordering pattern leads to a layer containing monomeric and dimeric units (Figures 1c and 3d) and AACC pattern to a layer containing only dimeric units (Figures 1b and 3e).

CdI$_2$ structure allows also for derivation of the layers with corner-sharing plaquettes, which may be achieved by ACAC ordering pattern and is in fact observed in low-temperature γ-PdCl$_2$ (Figure 1d, bottom) [38]. However, the predicted layered polymorphs of AgCl$_2$ with corner-sharing no longer belong to the CdI$_2$ family but rather to rutile and fluorite family as manifested by change of the axial coordination of the Ag atoms from intralayer to interlayer one (Figure 1d–e). Note that in all the ribbon and layered polymorphs derived from the CdI$_2$ structure, the cations are always octahedrally coordinated by intralayer anions; that is, by anions belonging to the same CdI$_2$-type layer. On the other hand, the silver atoms from the corner-shared AgCl$_2$ layers complete their octahedral coordination by axial chlorine atoms from adjacent layers (Figure 1). To do so, the CdI$_2$ layers must become less corrugated. Such geometrical arrangements are characteristic of a rutile structure. This prototypical structure consists of three-dimensional network of corner- and edge-shared octahedra. Orbital ordering at Jahn-Teller active cations in these octahedra may result in formation of layers consisting of corner-shared plaquettes, where each metal cation from each plaquette is axially coordinated by anions from adjacent layers, as exemplified by CuF$_2$ structure (Figure 4b). Furthermore, various stacking patterns of these layers may be realized. The simplest AA stacking is stabilized in the monoclinic CuF$_2$ type and the ABAB stacking in the orthorhombic AgF$_2$ type. Yet another structure with AB'AB' stacking was found that differs from the AgF$_2$ type by smaller relative shift of the layers (an intermediate between CuF$_2$ and AgF$_2$ structure) (Figure 5). These various stacking patters result in different axial contacts of the cations and diverse packing efficiencies. Recall that CuF$_2$ is rutile type structure. On the other hand, the AgF$_2$ structure is related to fluorite structure, where the cations reach for two additional ligands to complete a cubic coordination [11]. Compounds with rutile-like structures often transform to denser fluorite-like structure under pressure. Thus, three distinct stable layered forms of AgCl$_2$-CuF$_2$ type, AgF$_2$ type, and intermediate between the two with different stacking of the layers, might be achieved under different pressure conditions. More on that later.

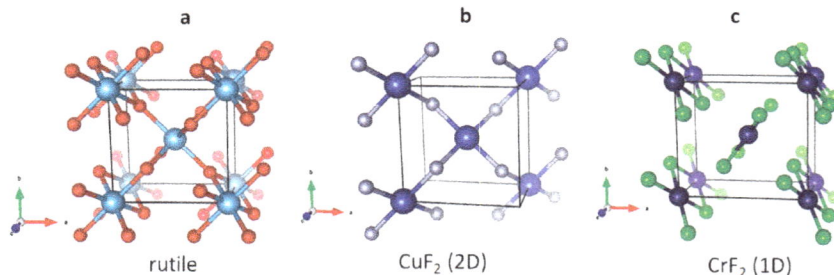

Figure 4. Crystal structure of rutile highlighting the edge- and corner-shared octahedra (**a**), rutile-derived two-dimensional (2D) CuF2- (**b**) and a hypothetical one-dimensional (1D) ribbon structure (**c**) achieved by octahedral elongation along two distinct octahedral directions.

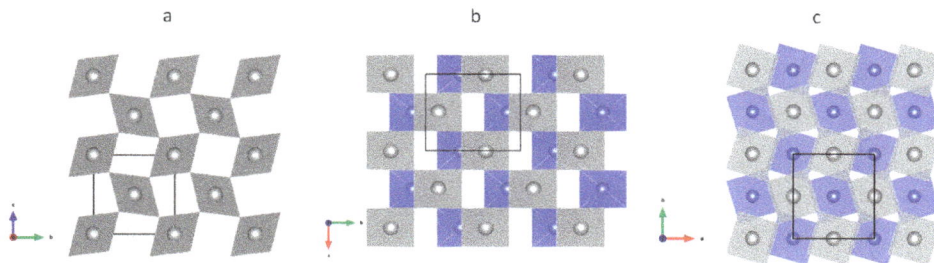

Figure 5. Three stacking patterns realized in the predicted layered $AgCl_2$ polymorphs with corner-sharing of $[AgCl_4]$ plaquettes: AA stacking of rutile-related CuF_2 type (**a**), AB' stacking (**b**), and AB stacking of fluorite-related AgF_2 type (**c**). The $[AgCl_4]$ plaquettes belonging to different layers are distinguished by different color (blue and grey).

Similarly, as in the case of the layered structures, several polytypes were found for the ribbon structures. While the $AgCl_2$ ribbons maintain the layered organization of the CdI_2 prototype with the interlayer contacts being longer than the intralayer ones, stacking of these layers may vary (Figure 6). We have obtained various polytypes in our search using evolutionary algorithms. The simplest stacking corresponds to one layer per unit cell that directly relates to the CdI_2 type structure. The original trigonal *P*-3m symmetry of the CdI_2 type is however lowered to triclinic due to the presence of the Jahn-Teller active Ag^{II} cation. Additionally, a monoclinic C2m structure of copper dihalide type was found in our scrutiny, as well as an orthorhombic and a triclinic version of the $PdCl_2$ polymorphs (Figure 6). Note, $PdCl_2$ crystallizes in two ribbon-like structures, a high-temperature orthorhombic and a low-temperature monoclinic form, which differ only in the monoclinic angle. Our results provide theoretical support for the observed strong tendency of the late TM dihalides with Jahn-Teller cations to form ribbon-like crystal structures exhibiting various packing.

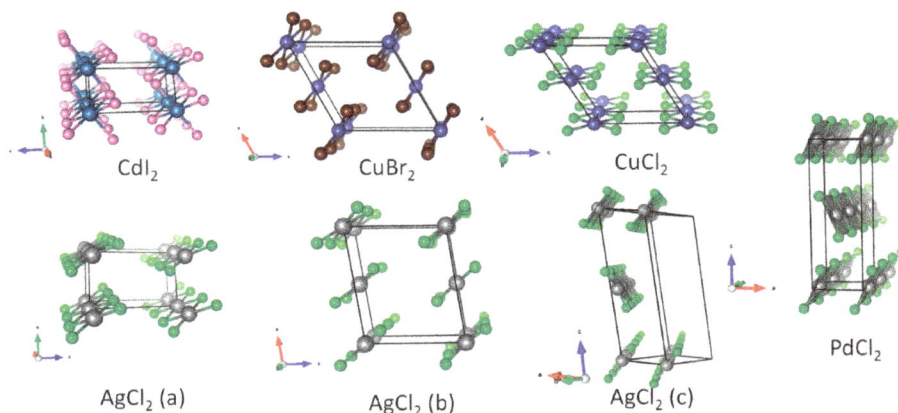

Figure 6. Three hypothetical $AgCl_2$ ribbon-like polytypes (bottom) obtained with evolutionary algorithms and their relation to the CdI_2 prototype (top left) and the known metal halides with ribbon structure (top and right). The differences of computed energy for (**a**–**c**) are minuscule.

Note that both rutile and fluorite structures are prototype structures for ionic crystals, while the CdI_2 structure is preferred by compounds forming more covalent bonds. In the predicted polymorphs of $AgCl_2$, we see frequent realization of structures related to both more ionic as well as more covalent structural types. This may be a manifestation of the intermediate character of the chemical bonding in $AgCl_2$.

3.2. The Unusual Ag(I)Cl(Cl$_2$)$_\frac{1}{2}$ Polymorph

As explained in the introduction, one of the key difficulties in preparation of the Ag(II) dichloride from elements or from AgCl and excess of Cl$_2$ is related to the fact that Ag(II) is a potent oxidizer. This means that silver might prefer to adopt its most common monovalent state (as AgCl), while the excess Cl atom would be forced to form Cl–Cl bonds with other similar species around. On the other hand, the so-formed Cl$_2$ is known to interact with Cl$^-$ anions in ionic compounds, by forming an asymmetric [Cl$^-$ ... Cl$_2$] or even symmetric [Cl-Cl-Cl$^-$] [39–41] trichloride anion. While the propensity of Cl$_3^-$ to form is much smaller than that of the related triiodide anion, yet such a Lewis structure should not escape our attention. Indeed, the XTalOpt quest has yielded one structure with the Ag(I)[Cl(Cl$_2$)$_\frac{1}{2}$] formulation (Figure 7). It consists of AgCl double layers with Cl$_2$ molecules sandwiched in between them.

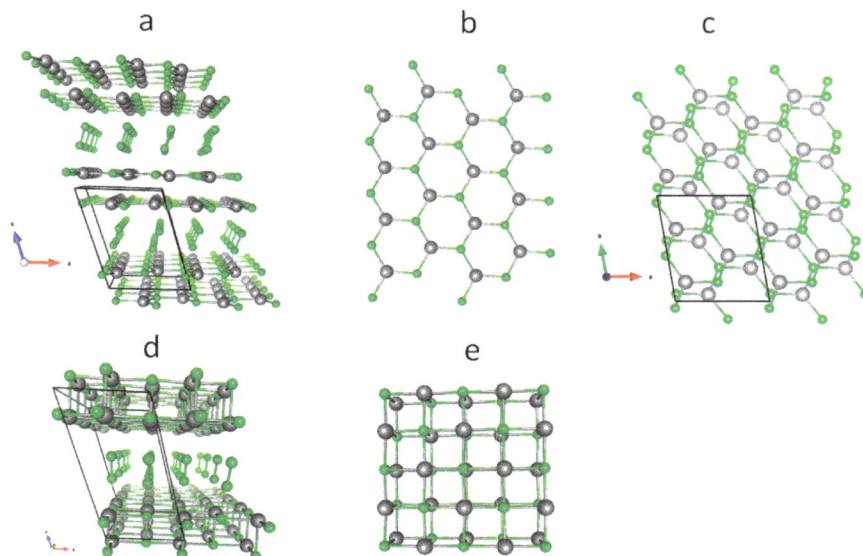

Figure 7. The predicted Ag(I)[Cl(Cl$_2$)$_\frac{1}{2}$] polymorphs: the crystal structure with hexagonal AgCl$_2$ double layers highlighting the unit cell, one AgCl layer and stacking of the AgCl layers, respectively (**a–c**), and an alternative structure with rock-salt AgCl double layers (**d,e**).

The structures of Ag(I)[Cl(Cl$_2$)$_\frac{1}{2}$] consist of the AgCl layers which can form either rock-salt layers or pseudo-hexagonal ones with only three short Ag-Cl bonds (Figure 7). The appearance of the rock salt layers seems natural since AgCl in rock salt structure is well known. However, the hexagonal layers are unknown among plain Ag(I) halides, with AgX (X = F, Cl, Br, I) adopting an ionic NaCl polytype (CN = 6), while AgI additionally takes on several structures with tetrahedral coordination of cation (CN = 4) (i.e., wurtzite, sphalerite, SiC(4H), etc.). In addition, CN of 7 is also possible for AgCl at rather low pressure of circa 1 GPa (TiI polytype [42]). The very low CN of 3 for Ag(I) in pseudo-hexagonal BN-like layer and a very short bond length of 2.521 Å indicates more covalent in respect the rock salt structure (six bonds at 2.773 Å [42]). Since the intra-sheet Ag-Cl bonding is covalent, it is not surprising to see that the interactions of Cl$_2$ with Cl$^-$ anions are far from symmetric, with intra-molecular Cl-Cl bond of 2.054 Å (slightly longer than that found for molecular solid of Cl$_2$, 1.97 Å), and Cl ... Cl$^-$ separation of 2.874 Å. That the Cl-Cl bond length is slightly longer than for free Cl$_2$ obviously stems from the donor-acceptor character of the Cl$^-$ ... Cl$_2$ interactions, and slight occupation of the sigma* orbital of Cl$_2$.

Formation of AgCl intercalated with Cl_2 molecules is peculiar given substantial lattice energy of AgCl solid, and little energy penalty to break weak Cl^- ... Cl_2 interactions upon phase separation to AgCl and $\frac{1}{2} Cl_2$. Their appearance in our quest is probably related to the limit imposed to Xtalopt on number of formula units and it marks the tendency towards the phase separation. We will turn to stability of these structures in the next section and discuss pressure effects further on.

3.3. Relative and Absolute Energetic and Thermodynamic Stability of Several Important Polymorphic Forms of AgCl₂

The crystal structures of all polymorphs considered here has been provided as cif files in Supplementary Materials (S2).

At DFT + U + vdW level the edge-sharing connectivity that leads to infinite $AgCl_2$ stripes (Figure 1e) was found to be the most energy preferred one among the $Ag(II)Cl_2$ polymorphs. All ribbon polytypes are maximally 5 meV/FU apart in energy. Among the layered structures, the most preferred is the ramsdellite related structure, then the γ-MnO_2 related (Figure 1b,c and Figure 2a,b) and finally the CuF_2 and AgF_2 related structures. The ramsdellite related structure (monoclinic space group) is only circa 40 meV/FU higher in energy than the ribbon polymorphs (Table 1). This energy order reflects preference of $Ag(II)Cl_2$ for edge connectivity of the $[Ag(II)Cl_4]$ square-planar units. Notably, all ribbon and puckered layered polymorphs are maximally 60 meV/FU apart. There is a considerable energy gap of about 200 meV/FU between the structures with puckered and flat layers; this is a manifestation of the fact that $Ag(II)$-Cl^- bonding is markedly covalent and it is characterized by close-to sp^2 hybridization at Cl atoms, which in turn comes with bending of the Ag–Cl–Ag angles. Concerning the structures containing Ag(I), the unusual $Ag(I)[Cl(Cl_2)_{\frac{1}{4}}]$ form with rock salt AgCl double layers is preferred over the one with hexagonal layers by 136 meV/FU. Furthermore, it represents the overall global minimum. The zero-point energy further plays in favor of this structure, by additional 10 meV/FU. Within the DFT + U picture, all predicted $AgCl_2$ polymorphs have negative formation energies and are thus energetically preferred over the elemental silver and molecular chlorine. However, they are metastable with respect to AgCl crystal. Calculated DFT + U energies of the lowest energy $AgCl_2$ forms are listed in Table 1 along with AgCl, molecular chlorine in its high-temperature polymorphic form [43], and elemental silver.

Table 1. Calculated DFT + U volumes and energies including van der Waals correction (E), zero-point energies (ZPE) and formation energies calculated in respect to the elemental crystals ($E^1_{form} = E - E_{Ag + Cl2}$) as well as to AgCl and $\frac{1}{2} Cl_2$ ($E^2_{form} = E - E_{AgCl + \frac{1}{2} Cl2}$) for five prototypical $AgCl_2$ polymorphs. The formation energies are calculated considering high-temperature crystal structure of Cl_2 [43]. Volume values in brackets come from experiment. The formation energy for AgCl is calculated as $E_{form} = E - E_{Ag + \frac{1}{2} Cl2}$. FU = formula unit.

Phase	Z	V/FU (Å³)	E (eV/FU)	E^1_{form}/FU (eV)	E^2_{form}/FU (eV)	ZPE/FU (eV)	V_{form}/FU (Å³)
Ribbon AgCl₂ (CdI₂ related)	4	65.7	−7.683	−0.840	+0.136	0.080	+3.6
Layered AgCl₂ (ramsdellite related)	4	66.7	−7.643	−0.800	+0.176	0.077	+4.6
Layered AgF₂ type	4	63.0	−7.625	−0.782	+0.194	0.081	+0.9
Ag(I)[Cl(Cl₂)$_{\frac{1}{4}}$] (rocksalt AgCl layers)	8	65.0	−7.718	−0.875	+0.101	0.070	+2.9
Ag(I)[Cl(Cl₂)$_{\frac{1}{4}}$] (hex AgCl layers)	8	77.2 *	−7.582	−0.739	+0.237	0.069	+15.1 *
AgCl + $\frac{1}{2}$ Cl₂		62.1	−7.819			0.027	
Cl₂	4	47.6 (58.1)	−4.288			0.029	
AgCl	4	38.3 (42.7)	−5.675	−0.976		0.021	−1.0
Ag fcc	4	15.5 (17.1)	−2.555			0.012	

* The unusually large calculated volume of this phase clearly suggests that it originates from attempts of XtalOpt to separate Cl₂ and AgCl phases, hence this phase may not correspond to any real local minimum, i.e., it may not be observable.

Inspection of the calculated energies and volumes of various phases of the $AgCl_2$ stoichiometry (Table 1) reveals that:

1. While all forms of $AgCl_2$ are stable with respect to elements, none of $AgCl_2$ polymorphs is energetically stable at T → 0 K and p → 0 atm with respect to products from Equation (3), i.e., AgCl and $\frac{1}{2}$ Cl_2.

2. The (relatively) most stable phase is that of $Ag(I)[Cl(Cl_2)_{\frac{1}{4}}]$ (rocksalt AgCl layers), as it falls at circa 0.1 eV above AgCl + $\frac{1}{2}$ Cl_2.

3. The ZPE correction changes very little the relative ranking of structures (it varies by no more than 12 meV for various phases), and for absolute stability of phases with respect to products (it destabilizes them by additional circa 42–53 meV), as could be expected for the system composed of rather heavy elements, Ag and Cl.

We have recalculated the total electronic energies and volumes of selected polymorphs also on much more resources-consuming hybrid DFT level using HSE06 functional (Table 2). Guided by the previous result, we did not perform daunting ZPE calculations this time.

Table 2. Calculated HSE06 volumes and energies and formation energies calculated with respect to the elemental crystals ($E^1_{form} = E - E_{Ag + Cl2}$) as well as to AgCl and $\frac{1}{2}$ Cl_2 ($E^2_{form} = E - E_{AgCl + \frac{1}{2} Cl2}$) for five prototypical $AgCl_2$ polymorphs. The formation energies are calculated considering high-temperature crystal structure of Cl_2 [43]. Volume values in brackets come from experiment. The formation energy for AgCl is calculated as $E_{form} = E - E_{Ag + \frac{1}{2} Cl2}$.

Phase	Z	V/FU (Å³)	E (eV/FU)	E¹form/FU (eV)	E²form/FU (eV)	ZPE/FU (eV)	Vform/FU (Å³)
Ribbon $AgCl_2$ (CdI₂ related)	4	77.6	−9.848	−1.030	+0.052	ND	+8.1
Layered AgF_2 type	4	68.7	−9.727	−0.909	+0.173	ND	−0.8
$Ag(I)[Cl(Cl_2)_{\frac{1}{4}}]$ (rocksalt AgCl layers)	8	70.7	−9.839	−1.021	+0.061	ND	+1.2
$Ag(I)[Cl(Cl_2)_{\frac{1}{4}}]$ (hex AgCl layers)	8	83.9 *	−9.837	−1.019	+0.063	ND	+14.4 *
$AuCl_2$-type (disproportionated)	4	76.9	−9.736	−0.918	+0.164	ND	+8.2
AgCl + $\frac{1}{2}$ Cl_2		69.5	−9.900			ND	
Cl_2	4	56.8 (58.1)	−5.536			ND	
AgCl	4	41.1 (42.7)	−7.132	−1.082		ND	−4.2
Ag fcc	4	16.9 (17.1)	−3.282			ND	

* The unusually large calculated volume of this phase clearly suggests that it originates from attempts of XtalOpt to separate Cl_2 and AgCl phases, hence this phase may not correspond to any real local minimum, i.e., not be observable.

The hybrid DFT results for $AgCl_2$ (Table 2) show that:

1. While all forms of $AgCl_2$ are stable with respect to elements, none of $AgCl_2$ polymorphs is energetically stable at T → 0 K and p → 0 atm with respect to products from Equation (3), i.e., AgCl and $\frac{1}{2}$ Cl_2; thus, confirming the DFT + U + vdW (van der Waals correction) results.

2. The (relatively) most stable phase is that of ribbon $Ag(II)Cl_2$ form as it falls at a mere 52 meV above AgCl + $\frac{1}{2}$ Cl_2.

3. HSE06 calculations predict the unit cell volumes of Ag, Cl_2, and AgCl quite well. The large calculated volume of the ribbon polymorph should be taken with a grain of salt, and this structure is bound only by weak vdW inter-ribbon interactions. The layered AgF_2-type structure is the only one for which the formation reaction volume is slightly negative.

Here, the unusual $Ag(I)[Cl(Cl_2)_{\frac{1}{4}}]$ form with hexagonal AgCl double layers and its rock salt layer analogue were found to be energetically almost degenerate within 2 meV/FU. The unusual $Ag(I)[Cl(Cl_2)_{\frac{1}{4}}]$ form with hexagonal layers was found to be only 11 meV/FU higher in energy in respect to the ribbon polymorph. Recall that the ZPE of the $Ag(I)[Cl(Cl_2)_{\frac{1}{4}}]$ forms is by circa 10–11 eV/FU lower with respect to the ribbon polymorph (Table 1), which points to factual energy degeneracy of all three solutions considering the hybrid DFT free energies and DFT + U ZPE energies.

Hybrid DFT was also used to model mixed valence (i.e., charge density wave) $Ag(I)Ag(III)Cl_4$ solution, which could not be captured properly on DFT + U level. We have chosen for this purpose crystal structure of $AuCl_2$, which forms molecular crystal with weakly bonded $Au(I)_2Au(III)_2Cl_8$

units [12]. Ag(III) cations are here in square planar [AuCl$_4$] coordination and Au(I) in linear [AuCl$_2$] coordination. These molecular units are stacked along one direction along which they polymerize into infinite chains under Au → Ag substitution. In the polymerized chains, the Ag(III) cations retain the square-planar coordination, while the Ag(I) cations pick up third chlorine ligand to form triangular instead of linear coordination. The triangular coordination is a consequence of Ag(I) moving closer to a chlorine atom belonging to the Ag(III) from the neighboring Ag(I)Ag(III)Cl$_2$ molecular unit. The Ag(I)-Cl bonds are then obviously longer (2.5 Å) in comparison to the Au(I)-Cl ones (2.3 Å) in the original AuCl$_2$ structure. On DFT + U level, the model converges to the one featuring chains of the comproportionated cations (Ag^{I}Ag^{III} → Ag^{II}Ag^{II}). This comproportionation is structurally manifested by Ag(I) cation picking up a fourth chlorine atom with which it completes square planar coordination of newly formed Ag(II) cation (the newly created Ag-Cl bond is highlighted by red dashed line in Figure 8b, bottom). Such polymerized Ag(II)Cl$_2$ chains are isostructural with recently discovered tubular form of AgF$_2$ that forms under high pressure (Figure 8c,d) [11]. In AgCl$_2$, the mixed valence chains are slightly energetically preferred (by circa 10 meV/FU) over the comproportionated ones at the hybrid DFT level. However, both are 100 meV/FU higher in energy with respect to the lowest energy ribbon polymorph.

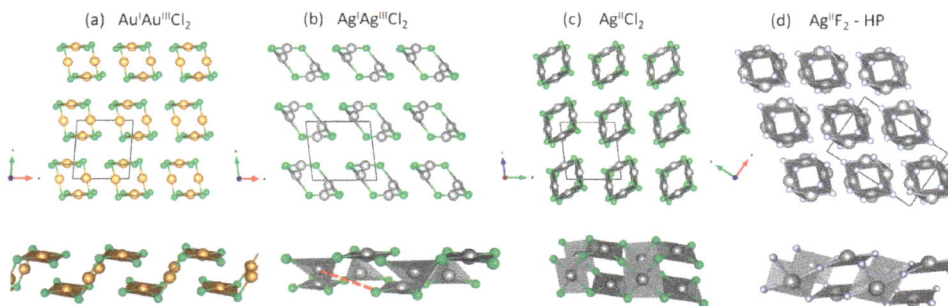

Figure 8. Crystal structure of Au(I)Au(III)Cl$_4$ (**a**), AgCl$_2$ optimized in the Au(I)Au(III)Cl$_4$ structure (**b**), AgCl$_2$ optimized in the Au(I)Au(III)Cl$_4$ structure with the symmetry-enforced comproportionation Ag^{I}Ag^{III} → Ag^{II}Ag^{II} (**c**) and high-pressure polymorph of AgF$_2$ (**d**). Top view: stacking of the chains, bottom view: connectivity within single chain.

3.4. Impact of Temperature and Pressure on Stability and Polymorphism of AgCl$_2$

Due to very similar energies of different polymorphs of AgCl$_2$ at T → 0 K and p → 0 GPa (also at the HSE06 level), and relatively small energy favouring the products of Equation (3) (AgCl and $\frac{1}{2}$ Cl$_2$), stability and polymorphism of AgCl$_2$ are expected to be dependent on (p, T) conditions. Here, we look briefly at the impact of external parameters on stability of AgCl$_2$.

The influence of temperature on stability of AgCl$_2$ is expected to be small in the range where Cl$_2$ is solid or liquid (i.e., up to its boiling point of −34 °C); the large reaction volume for the ribbon polymorph (Table 2), which is overestimated anyway, is insufficient to stabilize this phase via entropy factor [44]. Further increase of temperature will lead to preference for AgCl + $\frac{1}{2}$ Cl$_2$ via the entropy (ST) factor of the Cl$_2$ gas. The ST factor for $\frac{1}{2}$ Cl$_2$ at 300 K equals 347 meV [45] and thus, assuming that most of reaction volume change corresponds to the volume of Cl$_2$ gas released, it may be estimated that delta G^0 of AgCl$_2$ formation is about + 0.4 eV at 300 K. While this is only 40% of what Morris predicted (i.e., circa 1 eV) [2], the value is still substantial. Our results point out at the lack of thermodynamic stability of AgCl$_2$ at any temperature conditions (in the absence of external pressure effects).

The situation is somewhat different when the impact of external pressure is considered. Here, the infinite-sheet AgF$_2$-like form could potentially be stabilized at elevated pressure, as its formation from solid AgCl and $\frac{1}{2}$ Cl$_2$ is accompanied by small volume drop. The common tangent method [17,46] allows for a rather crude estimate for the formation pressure of AgCl$_2$ of 35 GPa (at T → 0 K), and likely even higher pressures at elevated temperature. The more precise estimate requires calculations in the function of pressure to be performed, also including the ribbon polymorph, which should exhibit substantial compressibility, and several viable high-pressure polymorphs [11,47,48]. Moreover, while drawing the computed volume-based conclusions one should always remember that despite great performance of HSE06 functional for describing crystal and electronic structure of solids, the reproduction of van der Waals interactions is still imperfect. And since they tend to collapse fast under even moderate pressures, it could be that other polymorphic forms, such as the ribbon one, would become competitive at rather low pressures, even preceding the transformation to the layered form. The previously documented pressure-induced transformations of CuF_2 [47] and AgF_2 [11] as well as a large body of data for transition metal difluorides and dichlorides (see also [48] and references therein) seem to suggest this scenario as a viable one.

3.5. Magnetic Properties of Selected Polymorphic Forms of $AgCl_2$

If chemistry teaches us something important, it is that virtually any chemical composition may be studied in its metastable form, given that the local minimum is protected by sizeable energy and/or entropy barriers. Thus, while $AgCl_2$ may not be thermodynamically stable at a broad range of (p,T) conditions, it is still insightful to theoretically study selected properties of $AgCl_2$, and compare them to those of the related halides ($CuCl_2$, AgF_2, $AuCl_2$, etc.).

For all $Ag(II)Cl_2$ forms featuring paramagnetic silver, the magnetic ordering is of interest, especially that magnetic properties of Ag(II) fluorides are now under intense scrutiny [49,50]. Thus, we have looked at spin ordering patterns, spin exchange pathways, as well as relevant superexchange constants for the ribbon and layered polymorphs of $AgCl_2$ (Table 3).

Table 3. Calculated energies, relative energies, and magnetic moments on atoms for the ribbon and AgF_2-type model of $AgCl_2$ as calculated using DFT + U + vdW. The unit cell was optimized for the ground state model (AFM1 in a ribbon and AFM in AgF_2-type structure), while energies of the remaining magnetic models were calculated as single point energies from the ground state.

Form	Ordering Pattern	E/FU (eV)	E_rel/FU (meV)	Ag (m_B)	Cl(m_B)
Ribbon polymorph	AFM1 (AABB)	−7.683	0	±0.31	±0.21 for F joining two like Ag spins ±0.00 between opposite Ag spins
	AFM2 (ABAB)	−7.652	31	±0.31	±0.10 on all F
	FM (AAAA)	−7.640	43	+0.35	+0.26 on all F
AgF_2-type	AFM	−7.625	0	±0.24	±0.10
	FM	−7.549	76	+0.35	+0.26

Not surprisingly, in the case of the ribbon polymorph the magnetic ground state found here is identical to that exhibited by structurally related frustrated Heisenberg chain system, $CuCl_2$, i.e., the spin pattern is AABB [51]. Correspondingly, as for $CuCl_2$ we consider the next neighbor (J_1) as well as the next near neighbor (J_2) superexchange constants (Figure 9), while neglecting all weaker magnetic interactions [51]. From the equations relating the energies of AFM1, AFM2, and FM states, and using the same Hamiltonian as authors [51]:

$$E_{AFM1} = (+2\,J_1) \times N^2/4 + \text{constant (this is AFM5 or AFM4 model in [51], since weaker interactions are omitted)}$$

$$E_{AFM2} = (+2\,J_1 - 2\,J_2) \times N^2/4 + \text{constant (this is AFM2 or AFM3 model in [51], since weaker interactions are omitted)}$$

$$E_{FM} = (-2\,J_1 - 2\,J_2) \times N^2/4 + \text{constant (this is FM model in [51], since weaker interactions are omitted)}$$

where N is the number of unpaired spins per spin site (in the present case, N = 1), one may derive J_1 = −12 meV and J_2 = −62 meV. The respective values for $CuCl_2$ calculated with U = 7 eV for Cu [51], are: J_1 = +18.4 meV and J_2 = −24.5 meV. Our results indicate that—like for $CuCl_2$—$|J_2| > |J_1|$ and the spin-exchange interactions are geometrically frustrated (Figure 9). Interestingly, however, J_1 is antiferromagnetic for $AgCl_2$ while ferromagnetic for $CuCl_2$. This result probably stems from the fact that antiferromagnetic next neighbor ordering implies null magnetic moments on bridging two Cl atoms, while the FM one introduces very large moments on chloride bridges (Table 3). The former is preferred, as elements which are typical nonmetals (here, in the form of a formally a closed shell Cl^- anion) do not support spin density on them, since it implies breaking of the stable electronic octet. Indeed, the spin density calculated for $AgCl_2$ in ribbon form suggests that spin density on one type of Cl atoms is as large as 2/3 of that on silver sites. While this could be expected based on previous studies of Ag(II) in chloride host lattices [52], this factor certainly contributes to lack of stability of $AgCl_2$. After all, if most spin sits on Cl atoms, Cl radical tend to pair up and eliminate Cl_2 molecules. This is indeed what one sees when comparing the energy of polymorphic forms of $AgCl_2$ with respect to phase separated $AgCl + \frac{1}{2} Cl_2$. The situation found for $CuCl_2$ is much different, where the total magnetic moment of circa $0.5\mu_B$ sits mostly on copper site [51].

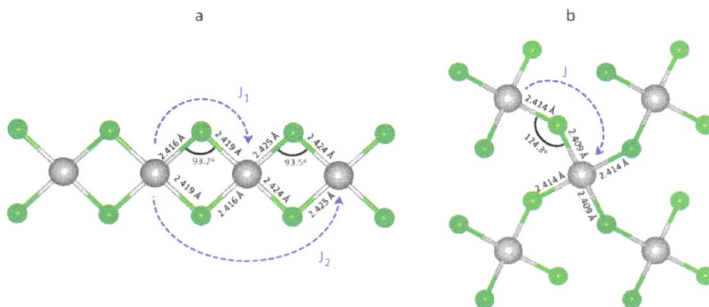

Figure 9. Fragment of the $AgCl_2$ ribbon with two relevant intra-chain superexchange constants considered in this study (**a**). Fragment of the corrugated $AgCl_2$ sheet with one relevant intra-sheet superexchange constant considered here (**b**).

Let us now scrutinize the magnetic interactions in the layered $AgCl_2$ polymorph (Table 3, Figure 9).

Here, four identical superexchange pathways link each Ag(II) site to its neighbors, as characterized by intra-sheet superexchange constant, J (the much weaker inter-sheet one will be omitted here). The ground state magnetic model corresponds to the familiar two-dimensional (2D) AFM ordering of spins, assumed also by AgF_2. Consequently, a spin flip to the FM state costs $(-4 J) \times N^2/4$, where N = 1. From the energy difference between the AFM and FM solutions we may extract J = −76 meV. For comparison the J found for AgF_2 at ambient conditions is −70 meV [53]. This implies a somewhat stronger magnetic superexchange for the Ag–Cl–Ag bridges than for the Ag–F–Ag ones, as indeed could be anticipated from the increased covalence of chemical bonding (Ag–Cl > Ag–F). This effect is, however, partially diminished by the Ag–Cl–Ag bridges being more bent (124 deg) than their Ag–F–Ag analogues found for AgF_2 (130 deg), and that decreases J for the former system [54], according to the Goodenough-Kanamori rules [55]. Corrugation of the sheets and departure of the Ag–Cl–Ag angle from 180 deg also results in the appearance of the magnetic moment of circa $0.1\mu_B$ at Cl atoms. This is half of what is found for the ribbon polymorph, yet still substantial, and must be viewed as a factor which contributes to the lack of stability of $AgCl_2$ with respect to elimination of Cl_2.

3.6. Electronic Properties of Selected Polymorphic Forms of $AgCl_2$

Having looked at magnetic properties, let us now examine electronic Density of States (DOS) and atomic (partial) DOS for four distinct polymorphs of $AgCl_2$ (Figure 10).

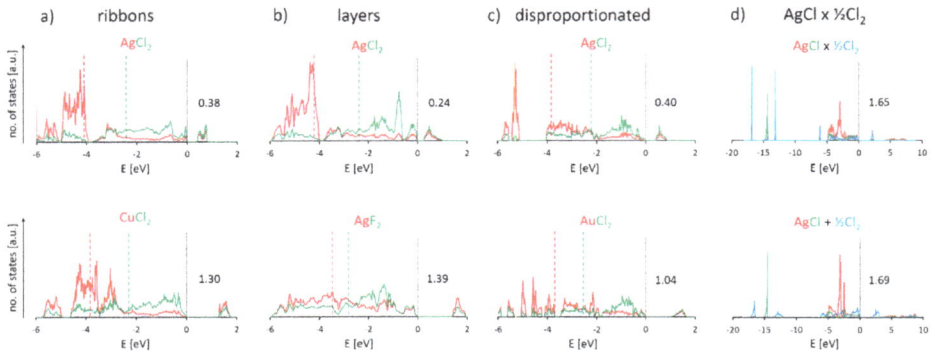

Figure 10. Comparison of the electronic density of states (DOS) graphs of four studied structures of $AgCl_2$ and their counterparts: (**a**) ribbon structure and $CuCl_2$, (**b**) layered structure and AgF_2, (**c**) disproportionated structure and $AuCl_2$, (**d**) $Ag(I)[Cl(Cl_2)_{\frac{1}{2}}]$ form with hexagonal $AgCl$ layers and the sum of eDOS of rocksalt $AgCl$ and solid Cl_2. Red and green dashed lines indicate the DOS-weighted average position of Cu/Ag/Au d and Cl/F p bands, respectively. The number above Fermi level in each graph indicates the fundamental band gap.

A glance at DOS graphs shows that all predicted polymorphs of $AgCl_2$ were found to have an insulating band gap. However, the calculated band gap at Fermi level tends to be substantially narrower in studied polymorphs of $AgCl_2$ than in their structural prototypes containing either a different group 11 metal ($CuCl_2$, $AuCl_2$) or a different halogen (AgF_2). In the case of $CuCl_2$-like ribbon structure, Ag 4d bands lie comparatively lower in energy than Cu 3d bands and are further separated from occupied Cl 2p states, which is in agreement with the stronger oxidizing properties of Ag(II) species as compared to Cu(II). The picture is somewhat similar in the layered structure: again, the Ag 4d states in $AgCl_2$ lie at higher binding energies and are more separated from Cl 2p states than in AgF_2, where the admixing between Ag 4d and F 1p states in AgF_2 is already substantial [53]. The same applies to the disproportionated form of $AgCl_2$ (i.e., an $AuCl_2$ polytype) as compared to its gold(II) analogue. The fact that Ag states are placed deeply below the Cl ones clearly contributes to the lack of stability of $AgCl_2$ in all polytypes, as oxidation of Cl^- anions by Ag(II) (or Ag(III) in disproportionated form) is facile. This is also reflected in very narrow fundamental bandgaps, which range between a mere 0.24 eV and 0.40 eV. The Maximum Hardness Principle (from Pearson [56]) dictates the preference for much larger bandgap calculated for $AgCl + \frac{1}{2}$ (Cl_2) (1.69 eV) and thus to a redox reaction.

As for the $AgCl_2$ polymorph consisting of hexagonal double layers of $AgCl$ interspersed with layers of Cl_2 molecules, we compared its electronic structure with the combination of eDOS of rocksalt-type $AgCl$ and solid chlorine (HT polymorph). Contributions from in-layer Cl atoms and from Cl_2 molecules between layers in this $AgCl_2$ polymorph are also plotted separately. The most apparent difference between otherwise similar graphs is that the bands pertaining to Cl_2 molecules are much sharper in $AgCl_2$ than in solid Cl_2, which indicates that there is relatively little bonding between them and $AgCl$ layers. On the other hand, Ag 4d bands in $AgCl$ layers of $AgCl_2$ are somewhat more diffuse than in rocksalt $AgCl$, which points to a slightly different (more covalent) Ag-Cl bonding character in hexagonal [$AgCl$] sublattice of $AgCl_2$ (as discussed in structural section above) than in ionic $AgCl$. In addition, the average position of Ag 4d band in this polymorph is circa −3 eV, which is 0.5–1.0 eV higher than in the other three studied polymorphs; this obviously stems from the fact that Ag(I) is present here rather than Ag(II). In this case, the gap is formed between top of the hybridized Ag^+(d)/Cl^-(p) states and the sigma* states of the Cl_2 molecules.

As to the relatively most stable forms of $AgCl_2$, i.e., a ribbon and layered polymorph, their band gaps have charge-transfer character; however, in idealized charge-transfer magnetic insulator the gap is formed between occupied nonmetal states (valence band) and the upper Hubbard band on metal (conduction band). Here, there is so severe mixture of the Ag and Cl states, that the top of the "ligand" band is composed in about 1/3 from Ag states, while the conduction band from a nearly equal mixture of the Ag and Cl states.

4. Conclusions

Our theoretical study for $AgCl_2$ including outcome of evolutionary algorithm structure prediction method suggests that $AgCl_2$ is metastable with respect to $AgCl + \frac{1}{2} Cl_2$ at $p \to 0$ atm and $T \to 0$ K conditions. Still, the energy penalty which must be paid for its synthesis from these substrates is relatively small and of the order of 0.1–0.15 eV per formula unit. Thermodynamic stability is smaller at ambient (p,T) conditions and of the order of 0.4 eV per formula unit, due to entropy factor for Cl_2 gas (reaction product). Still, $AgCl_2$ is not as severely unstable as previously predicted by Morris [2]. If prepared using some non-equilibrium methods, $AgCl_2$ would be metastable as indicated by lack of imaginary phonon modes for the structures we have scrutinized. $AgCl_2$ constitutes a challenge for theoretical methods as it allows for diverse charge instabilities, as well as on the verge of decomposition to simpler phases. The most stable polymorphic form of $AgCl_2$, according to hybrid DFT (HSE06) calculations, is related to $CuCl_2$ type, and it consists of infinite $[AgCl_{4/2}]$ ribbons. The lowest energy magnetic pattern for this phase is of the AABB type, thus similar to the one shown by $CuCl_2$. More complex magnetic ordering, i.e., helical, is also possible, due to frustration of NN and NNN superexchange interactions.

Formation of $AgCl_2$ should be facilitated by use of external pressure, as indicated by extrapolation based on the common tangent method. The thermodynamically stable form at circa 35 GPa has crystal structure related to that of AgF_2; and, like AgF_2, it shows 2D AFM ordering in its ground state.

Having uncovered the chemical identity of the most stable form of $AgCl_2$ together with its presumed magnetic properties, we may construct a simple table which demonstrates huge difference between coinage group metals [55,57] in terms of their real and hypothetical difluorides and dichlorides (Table 4). Thus, copper, silver, and gold are all different; indeed, the coinage metal group has been argued to contain the most dissimilar elements among all groups of the Periodic Table [58] and this is confirmed in our study of their dichlorides.

Table 4. Comparison of essential features (stability, structure) and magnetic properties (wherever applicable) of Group 11 difluorides and dichlorides, as seen from experiment and theoretical calculations.

Metal	Cu	Ag	Au
MF_2	Stable Layered 2D AFM [47]	Stable Layered 2D AFM [53]	Unstable Phase separation
MCl_2	Stable Ribbon 1D complex [51]	Metastable Phase separation [this work]	Stable Disproportionated Diamagnetic [12]

In conclusion, if prepared, $AgCl_2$ would be a very unusual metastable narrow-band gap (<0.4 eV) magnetic semiconductor. Quest for $AgCl_2$ should preferably utilize non-equilibrium, high-pressure, and low-T conditions. In the forthcoming work we will report our own attempts toward synthesis of $AgCl_2$ utilizing the diamond anvil cell setup.

Supplementary Materials: The following are available online at http://www.mdpi.com/2073-4352/9/8/423/s1, S1: Analysis of halogen-halogen interactions, S2: list of cif files.

Author Contributions: Conceptualization, W.G. and M.D.; methodology, M.D.; investigation, P.K., K.T. and A.G.; resources, W.G.; data curation, M.D.; writing—original draft preparation, M.D. and W.G.; writing—review and editing, M.D. and W.G.; visualization, M.D.; supervision, M.D. and W.G.; project administration, M.D. and W.G.; funding acquisition, M.D. and W.G.

Funding: Preliminary studies conducted a decade ago were performed within the TEAM project of the Foundation of Polish Science. Subsequent careful scrutiny was possible due to funding from the Polish National Science Center via Maestro project (UMO-2017/26/A/ST5/00570). AG's contribution was financed from Preludium project from NCN (2017/25/N/ST5/01976). MD and KT also acknowledge the Scientific Grant Agency of the Slovak Republic, grant No. VG 1/0223/19.

Acknowledgments: Most calculations were performed at the Interdisciplinary Center for Mathematical and Computational Modelling, the University of Warsaw, at Okeanos machine, within the project ADVANCE++ (GA76-19). MD and KT thank the Centre of operations of the Slovak Academy of Sciences for providing computational resources (supercomputer Aurel) within computation grant "Novel inorganic compounds from ab initio".

Conflicts of Interest: The authors declare no conflict of interest.

References

1. Fajans, K. Electronic Structure of Some Molecules and Crystals. *Ceramic Age* **1949**, *54*, 288.
2. Morris, D.F.C. The instability of some dihalides of copper and silver. *J. Phys. Chem. Solids* **1958**, *7*, 214–217. [CrossRef]
3. Rossini, F.D.; Wagman, D.D.; Evans, W.H.; Levine, S.; Jaffe, I. Selected values of chemical thermodynamic properties. Circular of National Bureau of Standards No. 500. 1952. Available online: https://archive.org/details/circularofbureau500ross/page/n7 (accessed on 14 August 2019).
4. Scatturin, V.; Bellon, P.L.; Zannetti, R. Planar coordination of the group IB elements: Crystal structure of Ag (II) oxide. *J. Inorg. Nucl. Chem.* **1958**, *8*, 462–467. [CrossRef]
5. Yvon, K.; Bezinge, A.; Tissot, P.; Fischer, P. Structure and magnetic properties of tetragonal silver(I, III) oxide, AgO. *J. Solid State Chem.* **1986**, *65*, 225–230. [CrossRef]
6. Hermann, A.; Derzsi, M.; Grochala, W.; Hoffmann, R. AuO: Evolving from dis- to comproportionation and back again. *Inorg. Chem.* **2016**, *55*, 1278–1286. [CrossRef] [PubMed]
7. Gammons, C.H.; Yu, Y.; Williams-Jones, A.E. The disproportionation of gold(I) chloride complexes at 25 to 200 C. *Geochim. Cosmochim. Acta* **1997**, *61*, 1971–1983. [CrossRef]
8. Malinowski, P.J.; Derzsi, M.; Gaweł, B.; Łasocha, W.; Jagličić, Z.; Mazej, Z.; Grochala, W. Ag^{II}SO4: A Genuine Sulfate of Divalent Silver with Anomalously Strong One-Dimensional Antiferromagnetic Interactions. *Angew. Chem. Int. Ed. Engl.* **2010**, *49*, 1683–1686. [CrossRef] [PubMed]
9. Elder, S.H.; Lucier, G.M.; Hollander, F.J.; Bartlett, N. Synthesis of Au(II) fluoro complexes and their structural and magnetic properties. *J. Am. Chem. Soc.* **1997**, *119*, 1020–1026. [CrossRef]
10. Derzsi, M.; Piekarz, P.; Grochala, W. Structures of late transition metal monoxides from Jahn-Teller instabilities in the rock salt lattice. *Phys. Rev. Lett.* **2014**, *113*, 025505. [CrossRef]
11. Grzelak, A.; Gawraczyński, J.; Jaroń, T.; Sommayazulu, M.; Derzsi, M.; Struzhkin, V.V.; Grochala, W. Persistence of Mixed and Non-intermediate Valence in the High-Pressure Structure of Silver(I,III) Oxide, AgO: A Combined Raman, X-ray Diffraction (XRD), and Density Functional Theory (DFT) Study. *Inorg. Chem.* **2017**, *56*, 5804–5812. [CrossRef]
12. Dell'amico, D.B.; Calderazzo, F.; Marchetti, F.; Merlino, S. Synthesis and molecular structure of [Au4Cl8], and the isolation of [Pt(CO)Cl5]$^-$ in thionyl chloride. *J. Chem. Soc., Dalton Trans.: Inorg. Chem.* **1982**, 2257–2260. [CrossRef]
13. Brückner, R.; Haller, H.; Steinhauer, S.; Müller, C.; Riedel, S. A 2D Polychloride Network Held Together by Halogen–Halogen Interactions. Angew. *Chem. Int. Ed. Engl.* **2015**, *54*, 15579–15583. [CrossRef] [PubMed]
14. Taraba, J.; Zak, Z. Diphenyldichlorophosphonium Trichloride–Chlorine Solvate 1:1, [PPh2Cl2]$^+$Cl3$^-$·Cl2: An Ionic Form of Diphenyltrichlorophosphorane. Crystal Structures of [PPh2Cl2]$^+$Cl3$^-$·Cl2 and [(PPh2Cl2)$^+$]2[InCl5]$^{2-}$. *Inorg. Chem.* **2003**, *42*, 3591–3594. [CrossRef] [PubMed]
15. Wickleder, M.S. AuSO4: A True Gold(II) Sulfate with an Au2^{4+} Ion. *Z. Anorg. Allg. Chem.* **2001**, *627*, 2112–2114. [CrossRef]
16. Derzsi, M.; Dymkowski, K.; Grochala, W. The Theoretical Quest for Sulfate of Ag^{2+}: Genuine Ag(II)SO4, Diamagnetic Ag(I)2S2O8, or Rather Mixed-Valence Ag(I)[Ag(III)(SO4)2]? *Inorg. Chem.* **2010**, *49*, 2735–2742. [CrossRef] [PubMed]
17. Kurzydłowski, D.; Grochala, W. Elusive AuF in the solid state as accessed via high pressure comproportionation. *Chem. Commun.* **2008**, 1073–1075. [CrossRef] [PubMed]

18. Kresse, G.; Hafner, J. Ab initio molecular dynamics for liquid metals. *Phys. Rev. B: Condens. Matter Mater. Phys.* **1993**, *47*, 558–561. [CrossRef] [PubMed]

19. Kresse, G.; Hafner, J. Ab initio molecular-dynamics simulation of the liquid-metal–amorphous-semiconductor transition in germanium. *Phys. Rev. B: Condens. Matter Mater. Phys.* **1994**, *49*, 14251–14269. [CrossRef]

20. Kresse, G.; Furthmüller, J. Efficiency of ab-initio total energy calculations for metals and semiconductors using a plane-wave basis set. *Comput. Mater. Sci.* **1996**, *6*, 15–50. [CrossRef]

21. Kresse, G.; Furthmüller, J. Efficient iterative schemes for ab initio total-energy calculations using a plane-wave basis set. *Phys. Rev. B: Condens. Matter Mater. Phys.* **1996**, *54*, 11169–11186. [CrossRef]

22. Kresse, G.; Joubert, D. From ultrasoft pseudopotentials to the projector augmented-wave method. *Phys. Rev. B: Condens. Matter Mater. Phys.* **1999**, *59*, 1758–1775. [CrossRef]

23. Parlinski, K. *PHONON Software*. 2010. Available online: http://wolf.ifj.edu.pl/phonon/ (accessed on 2 January 2008).

24. Zurek, E.; Grochala, W. Predicting crystal structures and properties of matter under extreme conditions via quantum mechanics: The pressure is on. *Phys. Chem. Chem. Phys.* **2015**, *17*, 2917–2934. [CrossRef] [PubMed]

25. Lonie, D.C.; Zurek, E. XtalOpt: An open-source evolutionary algorithm for crystal structure prediction. *Comput. Phys. Commun.* **2011**, *182*, 372–387. [CrossRef]

26. Zurek, E.; Hoffmann, R.; Ashcroft, N.W.; Oganov, A.; Lyakhov, A.O. A little bit of lithium does a lot for hydrogen. *Proc. Nat. Acad. Sci. USA* **2009**, *106*, 17640–17643. [CrossRef] [PubMed]

27. Baettig, P.; Zurek, E. Pressure-Stabilized Sodium Polyhydrides: NaH_n (n > 1). *Phys. Rev. Lett.* **2011**, *106*, 237002. [CrossRef]

28. Hooper, J.; Zurek, E. Rubidium Polyhydrides Under Pressure: Emergence of the Linear H_3^- Species. *Chem. Eur. J.* **2012**, *18*, 5013–5021. [CrossRef]

29. Liechtenstein, A.I.; Anisimov, V.I.; Zaane, J. Density-functional theory and strong interactions: Orbital ordering in Mott-Hubbard insulators. *Phys. Rev. B* **1995**, *52*, R5467. [CrossRef]

30. Kasinathan, D.; Kyker, A.B.; Singh, D.J. Origin of ferromagnetism in Cs_2AgF_4: The importance of Ag–F covalency. *Phys. Rev. B* **2006**, *73*, 214420. [CrossRef]

31. Grimme, S.; Antony, J.; Ehrlich, S.; Krieg, S. A consistent and accurate ab initio parametrization of density functional dispersion correction (DFT-D) for the 94 elements H-Pu. *J. Chem. Phys.* **2010**, *132*, 154104. [CrossRef]

32. Grochala, W.; Hoffmann, R. Real and Hypothetical Intermediate-Valence Ag^{II}/Ag^{III} and Ag^{II}/Ag^{I} Fluoride Systems as Potential Superconductors. *Angew. Chem. Int. Ed. Engl.* **2001**, *40*, 2742–2781. [CrossRef]

33. Grzelak, A.; Gawraczyński, J.; Jaroń, T.; Kurzydłowski, D.; Mazej, Z.; Leszczyński, P.J.; Prakapenka, V.B.; Derzsi, M.; Struzhkin, V.V.; Grochala, W. Metal fluoride nanotubes featuring square-planar building blocks in a high-pressure polymorph of AgF_2. *Dalton Trans.* **2017**, *46*, 14742–14745. [CrossRef] [PubMed]

34. Park, S.; Choi, Y.J.; Zhang, C.L.; Cheong, S.W. Ferroelectricity in an S = $\frac{1}{2}$ Chain Cuprate. *Phys. Rev. Lett.* **2007**, *98*, 057601. [CrossRef] [PubMed]

35. Gibson, B.J.; Kremer, R.K.; Prokofiev, A.V.; Assmus, W.; McIntyre, G.J. Incommensurate antiferromagnetic order in the S = 12 quantum chain compound $LiCuVO_4$. *Physica B* **2004**, *350*, e253. [CrossRef]

36. Yahia, H.B.; Shikano, M.; Tabuchi, M.; Kobayashi, H.; Avdeev, M.; Tan, T.T.; Liu, S.; Ling, C.D. Synthesis and characterization of the crystal and magnetic structures and properties of the hydroxyfluorides Fe(OH)F and Co(OH)F. *Inorg. Chem.* **2014**, *53*, 365–374. [CrossRef] [PubMed]

37. Hill, L.I.; Verbaere, A. On the structural defects in synthetic γ-MnO_2s. *J. Solid State Chem.* **2004**, *177*, 4706–4723. [CrossRef]

38. Evers, J.; Beck, W.; Göbel, M.; Jakob, S.; Mayer, P.; Oehlinger, G.; Rotter, M.; Klapötke, T. The Structures of δ-$PdCl_2$ and γ-$PdCl_2$: Phases with Negative Thermal Expansion in One Direction. *Angew. Chem. Int. Ed. Engl.* **2010**, *49*, 5677–5682. [CrossRef] [PubMed]

39. Brückner, R.; Haller, H.; Ellwanger, M.; Riedel, S. Polychloride Monoanions from $[Cl_3]^-$ to $[Cl_9]^-$: A Raman Spectroscopic and Quantum Chemical Investigation. *Chem. Eur. J.* **2012**, *18*, 5741–5747. [CrossRef]

40. Riedel, E.F.; Willett, R.D. NQR study of the trichloride ion. Evidence for three-center four-electron bonding. *J. Am. Chem. Soc.* **1975**, *97*, 701–704. [CrossRef]

41. Redeker, F.A.; Beckers, H.; Riedel, S. Matrix-isolation and comparative far-IR investigation of free linear $[Cl_3]^-$ and a series of alkali trichlorides. *Chem. Commun.* **2017**, *53*, 12958–12961. [CrossRef]

42. Hull, S.; Keen, D.A. Pressure-induced phase transitions in AgCl, AgBr, and AgI. *Phys. Rev. B* **1999**, *59*, 750–761. [CrossRef]
43. Donohue, J.; Goodman, S.H. Interatomic distances in solid chlorine. *Acta Cryst.* **1965**, *18*, 568–569. [CrossRef]
44. Jenkins, H.D.B.; Glasser, L. Volume-based thermodynamics: Estimations for 2:2 salts. *Inorg. Chem.* **2006**, *45*, 1754–1756. [CrossRef] [PubMed]
45. NIST. gov chemistry webbook database 2019. Available online: https://webbook.nist.gov/chemistry/ (accessed on 2 July 2019).
46. Horvath-Bordon, E.; Riedel, R.; Zerr, A.; McMillan, P.F.; Auffermann, G.; Prots, Y.; Bronger, W.; Kniep, R.; Kroll, P. High-pressure chemistry of nitride-based materials. *Chem. Soc. Rev.* **2006**, *35*, 987–1014. [CrossRef] [PubMed]
47. Kurzydłowski, D. The Jahn-Teller distortion at high pressure: The case of copper difluoride. *Crystals* **2018**, *8*, 140.
48. López-Moreno, S.; Romero, A.H.; Mejía-López, J.; Muñoz, A. First-principles study of pressure-induced structural phase transitions in MnF$_2$. *Phys. Chem. Chem. Phys.* **2016**, *18*, 33250–33263. [CrossRef] [PubMed]
49. Kurzydłowski, D.; Grochala, W. Prediction of extremely strong antiferromagnetic superexchange in silver (II) fluorides: Challenging the oxocuprates (II). *Angew. Chem. Int. Ed. Engl.* **2017**, *56*, 10114–10117. [CrossRef]
50. Kurzydłowski, D.; Grochala, W. Large exchange anisotropy in quasi-one-dimensional spin-$\frac{1}{2}$ fluoride antiferromagnets with a d(z^2)^1 ground state. *Phys. Rev. B* **2017**, *96*, 155140. [CrossRef]
51. Banks, M.G.; Kremer, R.K.; Hoch, C.; Simon, A.; Ouladdiaf, B.; Broto, J.-M.; Rakoto, H.; Lee, C.; Whangbo, M.-H. Magnetic ordering in the frustrated Heisenberg chain system cupric chloride CuCl$_2$. *Phys. Rev. B* **2009**, *80*, 024404. [CrossRef]
52. Aramburu, J.A.; Moreno, M. Bonding of Ag^{2+} in KCl lattice. *Solid State Commun.* **1986**, *58*, 305–309. [CrossRef]
53. Gawraczyński, J.; Kurzydłowski, D.; Ewings, R.A.; Bandaru, S.; Gadomski, W.; Mazej, Z.; Ruani, G.; Bergenti, I.; Jaroń, T.; Ozarowski, A.; et al. Silver route to cuprate analogs. *Proc. Nat. Acad. Sci. USA* **2019**, *116*, 1495–1500. [CrossRef]
54. Kurzydłowski, D.; Derzsi, M.; Barone, P.; Grzelak, A.; Struzhkin, V.V.; Lorenzana, J.; Grochala, W. Dramatic enhancement of spin–spin coupling and quenching of magnetic dimensionality in compressed silver difluoride. *Chem. Commun.* **2018**, *54*, 10252–10255. [CrossRef] [PubMed]
55. Pyykkö, P. Theoretical chemistry of gold. *Chem. Soc. Rev.* **2008**, *37*, 1967–1997. [CrossRef] [PubMed]
56. Pearson, R.G. Recent advances in the concept of hard and soft acids and bases. *J. Chem. Educ.* **1987**, *64*, 561–567. [CrossRef]
57. Schwerdtfeger, P. Relativistic effects in properties of gold. *Heteroatom Chem.* **2002**, *13*, 578–584. [CrossRef]
58. Grochala, W.; Mazej, Z. Unique silver (II) fluorides: The emerging electronic and magnetic materials. *Phil. Trans. A* **2015**, *373*, 20140179. [CrossRef] [PubMed]

crystals

MDPI

Article

Insight into the Optoelectronic and Thermoelectric Properties of Mn Doped ZnTe from First Principles Calculation

Wilayat Khan [1], Sikander Azam [2], Inam Ullah [3], Malika Rani [4], Ayesha Younus [5], Muhammad Irfan [3], Paweł Czaja [6] and Iwan V. Kityk [6,*]

[1] Department of Physics, Bacha Khan University, Charsadda 24420, KPK, Pakistan; walayat76@gmail.com
[2] Faculty of Engineering and Applied Sciences, Department of Physics, RIPHAH International University I-14 Campus, Islamabad 44000, Pakistan; sikander.physicst@gmail.com
[3] Department of Physics, The University of Lahore, Sargodha Campus 40100, Pakistan; 4inamullah@gmail.com (I.U.); bilalirfan104@gmail.com (M.I.)
[4] Condensed Matter Physics Lab, Department of Physics, The Women University, Multan 66000, Pakistan; dr.malikarani@yahoo.com
[5] Laser Matter Interaction and Nanosciences Lab, Department of physics, University of Agriculture Faisalabad, Faisalabad 38040, Pakistan; ayesha.younus@uaf.edu.pk
[6] Institute of Optoelectronics and Measuring Systems, Electrical Engineering Department, Czestochowa University of Technology, Armii Krajowej 17, 42-201 Czestochowa, Poland; czajap@el.pcz.czest.pl
*   Correspondence: iwank74@gmail.com

Received: 2 April 2019; Accepted: 7 May 2019; Published: 13 May 2019

Abstract: Using DFT band structure simulations together with semi-classical Boltzmann transport kinetics equations, we have explored the optoelectronic and transport features of $Mn_xZn_{1-x}Te$ (x = 8% and 16%) crystals. Optimization of the doping and related technological processes it is extremely important for optimization of the technological parameters. The Generalized Gradient Approximation is applied to compute the corresponding band structure parameters. We have applied the Generalized Gradient Approximation Plus U (GGA+U). We have demonstrated that $Mn_xZn_{1-x}Te$ (x = 8% and 16%) is a direct type band semiconductor with principal energy gap values equal to 2.20 and 2.0 eV for x = 8% and 16%, respectively. The energy gap demonstrates significant decrease with increasing Mn content. Additionally, the origin of the corresponding bands is explored from the electronic density of states. The optical dispersion functions are calculated from the spectra of dielectric function. The theoretical simulations performed unambiguously showed that the titled materials are simultaneously promising optoelectronic and thermoelectric devices. The theoretical simulations performed showed ways for amendment of their transport properties by replacement of particular ions.

Keywords: electronic properties; optical properties; thermoelectricity; semiconductors; electrical engineering

---

1. Introduction

It is well known that the pure ZnTe, crystals are direct type semiconductors (with E_g = 2.26 eV). More importantly, due to specific electronic and phonon features these materials have abundant potential for thermoelectric and optoelectronic applications [1–4]. Moreover, recently, ZnTe was doped by different transition metals that caused a huge attention due to their attractive magnetic properties, high fluorescence and structural compatibility with II–VI semiconductors and some essential III–V semiconductors, for example, GaAs [5–11]. In 2010, Mn-doped ZnTe was proposed due to its good theoretical band gap [12,13]. Thus, it can be used as a spintronic material and for studies of its magnetic properties.

For practical applications, the existence and optimization of the existing technology is a very hot topic. However, without reliable band structure parameters it is not possible to predict the ways in which the technology will further improve.

Addition of the transition metals opens a rare opportunity to vary their optoelectronic and thermal features in the desired directions. This is caused by a high flexibility of the 3d TM orbitals effectively interacting with the p orbitals of the anionic dopants. Many of them possess the localized d-states within the energy gap. Very interesting is use of ZnTe for the fabrication of green LED (light emitting diodes) [14] and various devices like THz emitters and detectors. The principal parameters are determined by the valence band state features originating from p Te states and their coordination by the cationic environment possessing space localized d-states. Today the main detectors in the visible spectral rage are based on Si. However, the band dispersion near the top of the valence band with respect to the bottom of conduction band is not very suitable for such applications. Conversely, the ZnTe has a suitable band energy gap and relatively high quantum efficiency. However, a strong density-functional theory (DFT) band structure calculation is absent, and most of the research in this direction are based on the oversimplified models without the band structure dispersion in k-space. Therefore, focusing on studies on ZnTe band dispersion and its changes after d-states doping is extremely important for the design of optoelectronic and thermoelectric devices of photodetectors [15].

Among a lot of technological approaches for thin-film ZnTe production, like molecular beam epitaxy, thermal vacuum evaporation, vapor phase epitaxy, physical vapor transport, hot wall evaporation, metallic–organic vapor phase epitaxy and electrodeposition [16–25], it is clear that the only way to amend these methods is knowledge about the principal changes of the hyperfine band structure after doping by 3d elements, and the origin of electronic states. In several works [26,27], the optoelectronic features of ZnX (X = O, S, Se, Te) have been studied within a framework of a DFT approach. However, the influence of doping has not been explored yet.

The goal of our manuscript is to study the effect of Mn impurities on optical and thermoelectric properties. Using first principles, and the total energy method, we have firstly applied the Generalized Gradient Approximation (GGA) in order to study the $Mn_xZn_{1-x}Te$ (x = 8% and 16%) band structure dispersion. It is shown that, indicates that GGA method is not enough get the exit gap therefore we have included a Hubbard-like contribution of the Mn d states, according to the so-called Generalized Gradient Approximation Plus U (GGA+U) is able to open the big band gap.

Following the reasons presented above in the present work we present studies of the band structure for the ZnTe single crystals doped by Mn using the first principle DFT method for the bulk material of $Mn_xZn_{1-x}Te$ (x = 8% and 16%), The results obtained may be very useful for the further materials engineering optimization of the technological doping processes to achieve enhanced optoelectronic and thermoelectric features.

2. DFT Calculation Technique

$Mn_xZn_{1-x}Te$ (x = 8% and 18%) possesses tetragonal symmetry. The basic lattice unit cell structures are depicted in Figure 1. We have applied the Full Potential Linear Augmented Plane wave method (FP-LAPW) method and at the beginning we optimized the initial structures within a framework of the one-electron WIEN2k package. GGA+U was used to evaluate the band electronic structure dispersion and the related optical function dispersion of the titled $Mn_xZn_{1-x}Te$ (x = 8% and 16%) materials.

We extended the primitive cell of ZnTe to 2 × 2 × 2 supercell containing 32 atoms (Zn and Te contains both 16 atoms), and used the supercell to simulate Mn doped systems. For the Mn doped ZnTe, one Zn atom substitutes for one/two (8/16 %) Zn, and denoted such systems as $Mn_xZn_{1-x}Te$ (x = 8% and 18%). For the $Mn_xZn_{1-x}Te$ (x = 8% and 18%) the muffin tin (MT) spheres radii were assumed to be equal to 1.8 and 2.0 Bohr, respectively. The values used for plane cutoff and harmonic expansion were l_{max} = 10 and $R_{MT}K_{max}$ = 8.0. The condition for the energy convergence in the self-consistent iteration was set to be equal to about 0.00001 eV. BoltzTrap code (which works with constant relaxation time) was used to calculate thermoelectric Seebeck coefficient, power factor, electrical and thermal conductivities using semi classical Boltzmann transport theory, based on the electronic band structure.

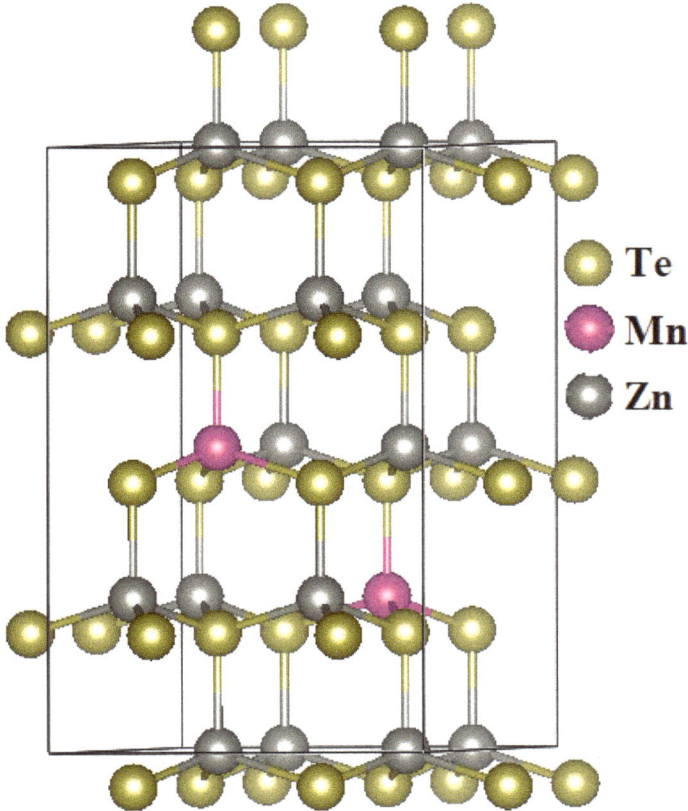

Figure 1. Representation and labeling of unit cell structure for $Mn_xZn_{1-x}Te$ (x = 8% and 16%) crystalline solid state alloys.

3. Results and Discussion

3.1. Electronic Structure

Using the modified Becke Johnson (mBJ) potential for high symmetry points [A→Γ→M→L→A→H→K→Γ] of the irreducible Brillouin zone (BZ), the electronic band energy diagrams for Up and Down states of $Mn_xZn_{1-x}Te$ (x = 8% and 16%) are depicted in the Figure 2. The spin-polarized calculations have been performed, in which the majority and minority spin electrons were treated separately. We applied the Generalized Gradient Approximation Plus U (GGA+U). Here the Fermi level is indicated by dashed line separating the valance band maximum and conduction band minimum.

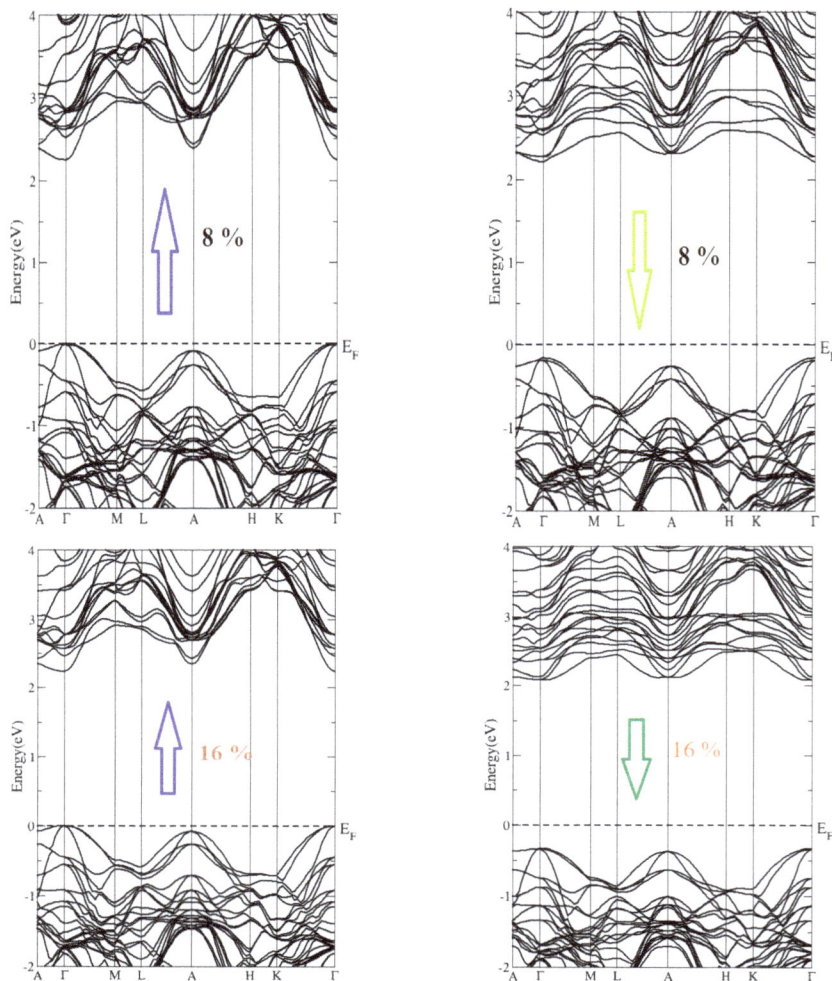

Figure 2. Electronic band structure along high symmetry points for $Mn_xZn_{1-x}Te$ (x = 8% and 16%).

It is obvious that these crystals have different band dispersion in the k-space. From the band structure, it was established that all the titled compounds were direct band semiconductors with energy band gap magnitudes equal to 2.20 and 2.0 eV for $Mn_xZn_{1-x}Te$ (x = 8% and 16%), respectively. These calculated band gap values were in sufficiently good agreement with the experimental values calculated from UV-vis diffuse reflectance spectra.

For all the band structures, principal contributions to the valance band minimum originated from 3d orbitals of Mn and p Zn orbitals, while the unoccupied states originated from 3p orbitals of S with a small admixture of other states of different atoms.

The spin polarized total density of states (TDOS) and partial density of states (PDOS) in a wide energy range is extended within −12.0 eV ~ 14.0 eV for the electronic states of $Mn_xZn_{1-x}Te$ (x = 8% and 16%) compounds depicted in the Figures 3 and 4. We saw only minor changes at the conduction band minimum for both percentages.

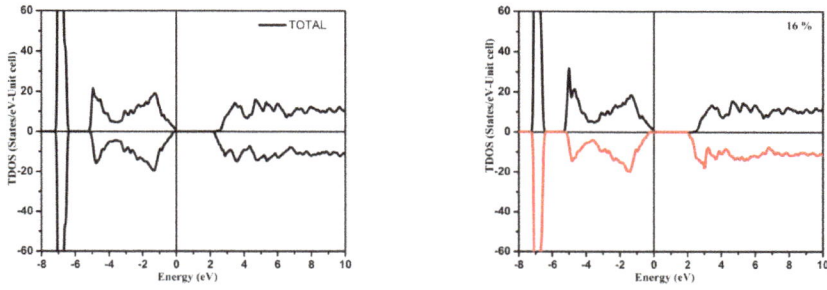

Figure 3. Total Density Up and Down of states for $Mn_xZn_{1-x}Te$ crystals (x = 8% and 16%).

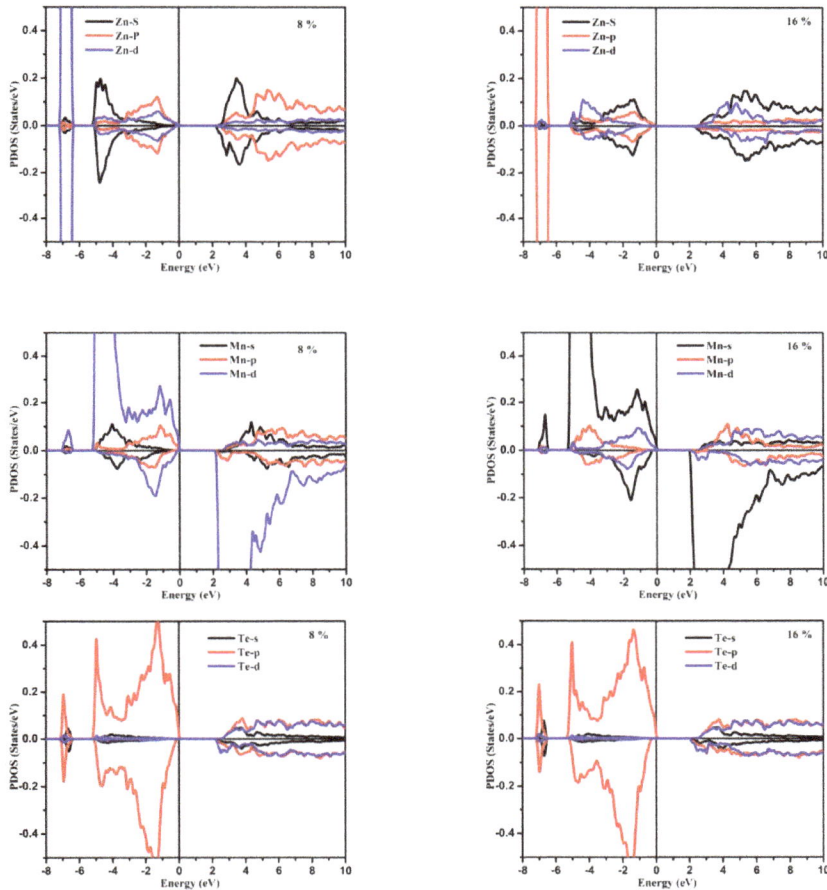

Figure 4. Up and Down states of the PDOS for $Mn_xZn_{1-x}Te$ crystals (x = 8% and 16%).

Following Figure 3, it is clear that for the 8% doped crystal the density of states (DOS) is comparable to the 16% doped crystals. Also, the energy separation between occupied and unoccupied states was established to be higher for 8% with respect to 16%.

The (PDOS) for the electronic states of $Mn_xZn_{1-x}Te$ (x = 8% and 16%) can be subdivided into three separate energy regions. For the compounds $Mn_xZn_{1-x}Te$ (x = 8% and 16%), a major contribution is s orbitals of Te with a small admixture coming from Zn s/p/d. For an energy interval extending within −7.2 … 6.5 eV, the PDOS, for the compounds Mn$_x$Zn$_{1-x}$Te (x = 8% and 18%) had a major contribution due to d and p originated Zn band states, along with the small contribution Te-p and Mn-s atoms.

In Mn, doped ZnTe Mn-s/p/d stats played a crucial role for reported energy range (−4.0–4.0 eV). A strong hybridization was observed between Zn s/p and d states along with the Mn s/p/d states. Its role in conduction bands became more prominent than the valence band.

3.2. Optical Function Dispersion

Study of the optical properties is important for understanding the electronic structure of the materials. These can be attained from the complex dielectric function $\varepsilon(\omega)$ which is expressed as

$$\varepsilon(\omega) = \varepsilon_1(\omega) + i\varepsilon_2(\omega) \tag{1}$$

The imaginary part $\varepsilon_2(\omega)$ is found from the momentum dipole transition matrix elements between the occupied and the unoccupied electronic states, and has been computed using Equation (1) [28],

$$\varepsilon_2^{ij}(\omega) = \frac{4\pi^2 e^2}{Vm^2\omega^2} \times \sum_{knn'\sigma} \langle kn\sigma|p_i|kn'\sigma\rangle\langle kn'\sigma|p_j|kn\sigma\rangle \times f_{kn}(1-f_{kn'})\sigma(E_{kn'}-_{kn}-\hbar\omega) \tag{2}$$

The dispersion of the real part of the dielectric function $\varepsilon(\omega)$ was computed using the imaginary part by using Kramer's- Kronig relations.

$$\varepsilon_1(\omega) = 1 + \frac{2}{\pi}P\int_0^\infty \frac{\omega'\varepsilon_2(\omega')}{\omega'^2 - \omega^2}d\omega' \tag{3}$$

The symbol P represents the principal value of the integral.

With the help of real and imaginary part dispersions for dielectric function, other optical properties were calculated. The complex index of refraction is written as

$$\tilde{n}(\omega) = n(\omega) + ik(\omega) \tag{4}$$

Here $n(\omega)$ is refractive index and $k(\omega)$ is extinction coefficient can be obtained from dielectric function.

$$n(\omega) = \left(\frac{\varepsilon_1(\omega) + \left(\varepsilon_1^2(\omega) + \varepsilon_2^2(\omega)\right)^{1/2}}{2}\right)^{1/2} \tag{5}$$

At low frequency (i.e., $\omega = 0$), the real part of the refractive index is called the static refractive coefficient $n_0 = [\varepsilon(0)]^{\frac{1}{2}}$.

So, from the complex dielectric function dispersion which contains both real and imaginary parts all the other optical functions were also calculated, such as absorption coefficient $I(\omega)$, energy loss function $L(\omega)$ and reflectivity $R(\omega)$. As the reflectivity is the percentage of reflected ray intensity on the incident ray intensity of electromagnetic waves on the system, it can be expressed as

$$R(\omega) = \frac{(n(\omega)-1)^2 - K(\omega)^2}{(n(\omega)+1)^2 - K(\omega)^2} \tag{6}$$

The absorption coefficient is the power absorbed in a unit length of solid, and is calculated by using this formula

$$I(\omega) = \frac{4\pi P(\omega)}{\lambda_0} \tag{7}$$

The energy loss function has been calculated as

$$L(\omega) = \frac{\varepsilon_2(\omega)}{\varepsilon_1^2(\omega) + \varepsilon_2^2(\omega)} \tag{8}$$

The optical absorption spectra may be considered an effective experimental tool to identify the hyperfine band electronic structure of crystalline solid state materials. At the beginning we explored the optical absorption spectra for $Mn_xZn_{1-x}Te$ (x = 8% and 18%). To describe the optical function dispersions of the titled crystals possessing tetragonal symmetry, only two tensor components ($\varepsilon^{xx}(\omega) = \varepsilon^{yy}(\omega)$ and $\varepsilon^{zz}(\omega)$) are sufficient following the general symmetry. But in the present work, we consider an average value of the two tensor components, both for the real part as well as in the imaginary part dispersion of the dielectric constant. The optical properties (the real ($\varepsilon_1^{ave}(\omega)$) and imaginary ($\varepsilon_2^{ave}(\omega)$) part of dielectric functions and other associated optical properties) have been investigated at the equilibrium constant at energies up to 25.0 eV and are illustrated in Figures 5 and 6. Real and imaginary parts of the dielectric function dispersion for $Mn_xZn_{1-x}Te$ (x = 8% and 18%) are shown in Figure 5.

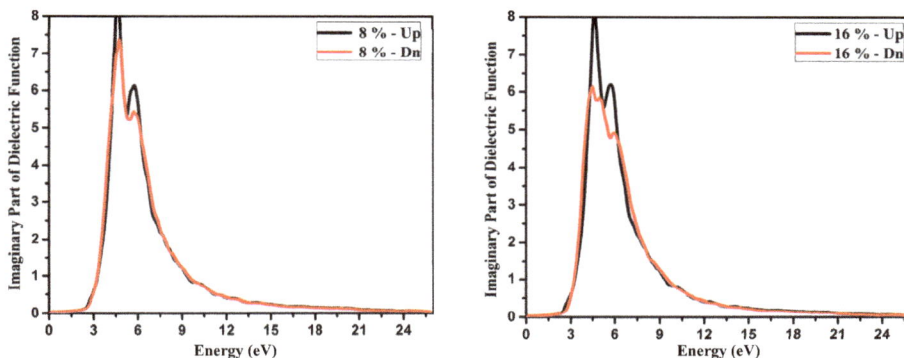

Figure 5. Calculated imaginary part (dark solid curve-black color for up spin) and (long solid curve-red color for down spin) spectra of $Mn_xZn_{1-x}Te$ (x = 8% and 16%).

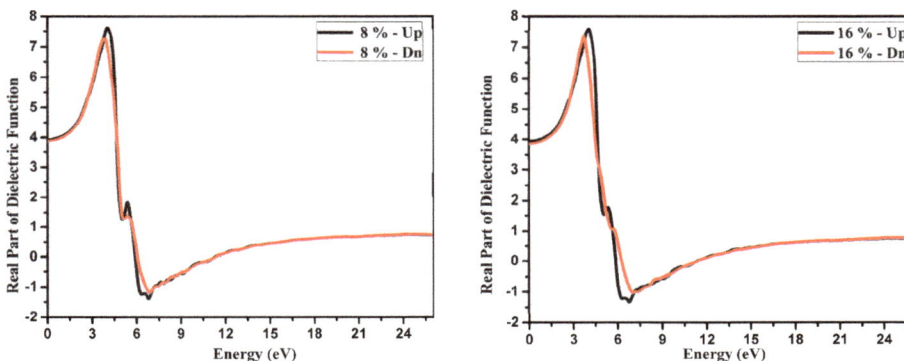

Figure 6. Calculated real part dispersion (dark solid curve-black color for up spin) and (long solid curve-red color for down) spectra of $Mn_xZn_{1-x}Te$ (x = 8% and 16%).

The determination of the imaginary part $\varepsilon_2^{ave}(\omega)$ confirms that the threshold energy of the dielectric function (i.e., first optical critical point) exists at energy equal to about at 2.6 eV for $Mn_xZn_{1-x}Te$ (x = 8% and 18%) compounds. This point is MV-MC splitting or ΓV-ΓC transition, which is in accordance with the threshold for direct optical inter-band transitions known as the fundamental absorption edge existing in the middle of the highest CBM and VBM BZ points. The spectra demonstrated a relatively fast increase in the fundamental absorption edge caused by the abrupt surpassing of many points which are contributing towards $\varepsilon_2^{ave}(\omega)$. The number of spectral peaks increased as the energy increased and we have detected the highest peaks at energy about 3.5 eV for $Mn_xZn_{1-x}Te$ (x = 8% and 18%). Weak anisotropy was noticed in spectra of $Mn_xZn_{1-x}Te$ (x = 8% and 18%) within the energy range situated between 3.2 ~ 7.0 eV. So, applying the Kramer-Kronieg relations [19,20], the real part dispersion is computed following the imaginary part dispersion (this is shown in Figure 6).

The principle spectral peaks of the real part, having a magnitude of < 7.0 for $Mn_xZn_{1-x}Te$ (x = 8% and 18%) compounds, were observed at energies about 3.5 eV. But the spectra curves were decreasing as the energy was increasing and crossed the zero line at about 6.0 eV for $Mn_xZn_{1-x}Te$ (x = 8% and 18%). The calculated values of static dielectric constant $\left(\varepsilon_1^{ave}(0)\right)$ were found to be equal to about 3.8 and 3.7 for $Mn_xZn_{1-x}Te$ (x = 8% and 18%) at the conditions of equilibrium lattice constant. We also studied the dispersions of associated optical constant like refractive index $n^{ave}(\omega)$, absorption spectra $I^{ave}(\omega)$, reflectivity $R^{ave}(\omega)$, and energy loss function $L^{ave}(\omega)$. So, following the results obtained it is clear that the Mn doping may effectively vary the resonance position of the spectral maxima for the titled crystals.

The principal method to find out how deeply light penetrates into the material is the determination of absorption coefficient $I(\omega)$ dispersion. The highly localized inter-band transitions mainly gave rise to absorption spectra. The calculated absorption coefficients dispersions are plotted in Figure 7. The spectra showed similar behaviors for the investigated materials, but also illustrated a few differences with significant optical anisotropy. The absorption edges were closely related to their band gaps and have been found to be equal to 2.5 eV for $Mn_xZn_{1-x}Te$ (x = 8% and 18%) compounds. In addition, as the energy increased, the absorption coefficient values obtained also were increasing and the maximum spectral peaks were situated in the spectral range situated within 5.0 ~ 9.0 eV. The spectra showed a sharp drop at 15.0 eV, which may be caused by electronic inter-band transitions, appearing only when a photon of specific energy is resonantly absorbed. The sharp decrease in the spectrum also shows a possibility of forbidden in dipole approached inter-band transitions in the band structure. Generally, it is necessary to have a direct band gap type for any crystalline material that may be considered as a promising candidate for photovoltaic, photoelectrical and even photo-thermal applications [20,21].

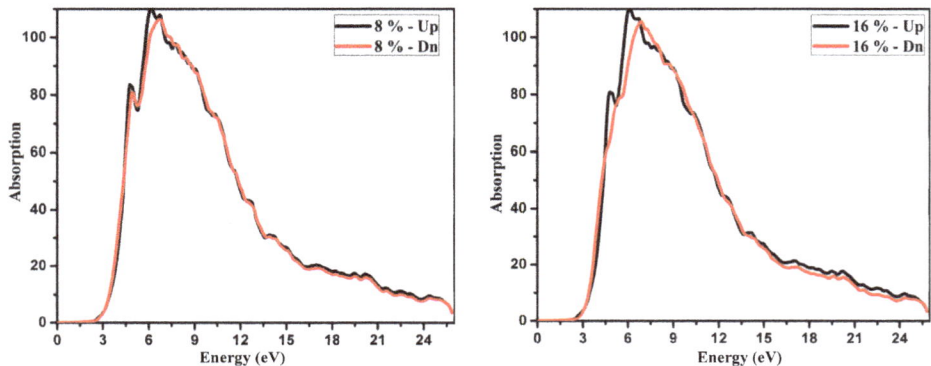

Figure 7. Calculated absorption spectrum (long solid curve-black color for up) and (long solid curve-red color for down) spectra for $Mn_xZn_{1-x}Te$ (x = 8% and 16%) crystals.

The refractive index n(ω) dispersion is shown in Figure 8. In this work we pay particular attention only to the average refractive index magnitudes. It is crucial there exists weak anisotropy between $Mn_xZn_{1-x}Te$ (x = 8% and 18%). In Figure 8, we have established that the highest refractive index peaks for $Mn_xZn_{1-x}Te$ (x = 8% and 18%) compounds are situated near the spectral energy equal to about 3.5 eV. After the energy limit mentioned, the titled peaks were substantially spectrally shifted toward lower energies as we moved from 8% to 16%, but as the energy increased from 3.5 eV a decrease in the spectral peaks was observed.

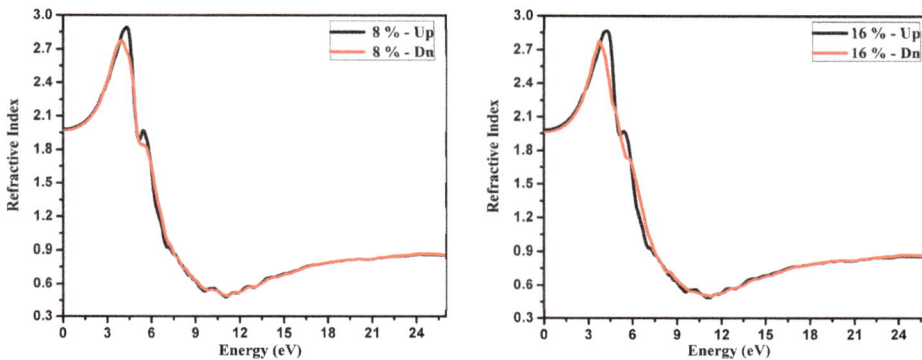

Figure 8. Calculated refractive index (long solid curve-black color for up) and (long solid curve-red color for down) spectra of $Mn_xZn_{1-x}Te$ (x = 8% and 16%) crystals.

Figure 9 shows that with the energy increases, we have an inverse relation with energy loss function (as shown in Figure 10), but it still shows minimal reflectivity in the visible region. It is also very important that at higher energies, all the three materials have a slightly higher reflectivity.

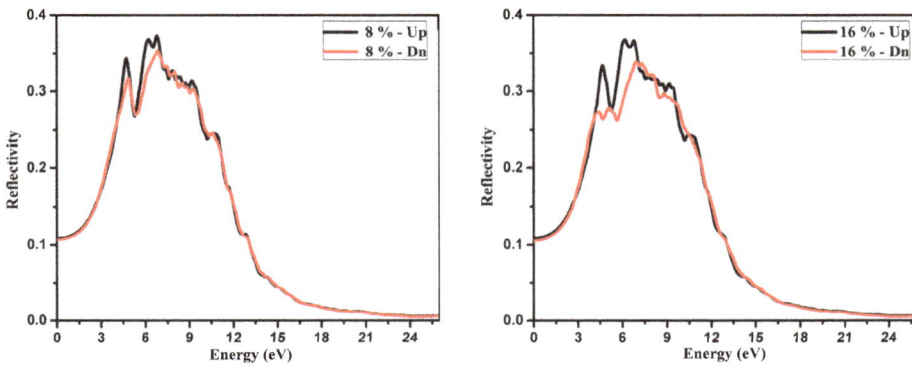

Figure 9. Calculated reflectivity (long solid curve-black color for up) and (long solid curve-red color for down) spectra of $Mn_xZn_{1-x}Te$ (x = 8% and 16%).

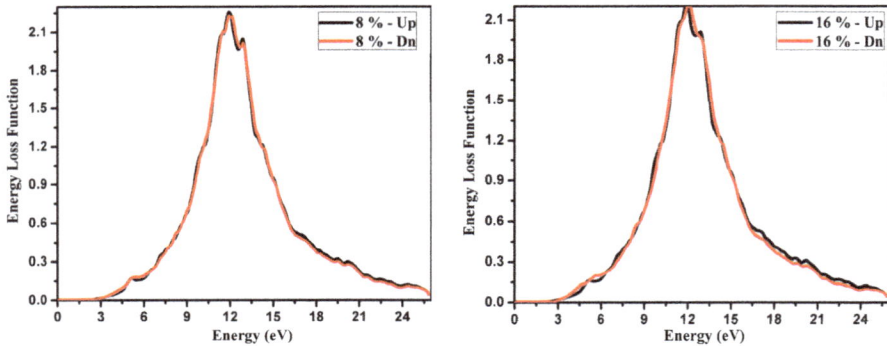

Figure 10. Calculated energy-loss spectrum (long solid curve-black color for up) and (long solid curve-red color for down) spectra for $Mn_xZn_{1-x}Te$ (x = 8% and 16%).

3.3. Thermoelectric Properties

We calculated and analyzed the electrical conductivity (σ), thermal conductivity (κ), Seebeck coefficient (S), power factor (PF), and figure of merit (ZT) as a function of temperature in the range 100–800 K. The average values of the electrical conductivity (σ^{ave}) were computed for three compounds, as shown in Figure 11. It is evident from the figure that all of the compounds showed different behavior with increasing temperature. In fact, average electrical conductivity (σ^{ave} (T)) of $Mn_xZn_{1-x}Te$ (x = 8 and 16%) is shown in Figure 11, and clearly demonstrates the fact that doping by 8% lead to a significant increase of electrical conductivity and temperature for spin up case, and for 8% spin down case we had a lower value of electrical conductivity obtained even at higher temperature. When the doping was enhanced up to 16% then there was a linear increase of conductivity versus electrical conductivity and temperature for up case, but at down case there was no change in electrical conductivity even at higher temperatures. These results may serve as an independent confirmation of the semi-conducting nature of these doped chalcogenides. At 100 K the values of σ^{ave} (T) for $Mn_xZn_{1-x}Te$ (x = 8% and 16%) considered were found to be equal to 0 and $0.45 \times 10^{18} (\Omega\ s)^{-1}$ for up and down cases, while for 16% the value of σ_{ave} (T) is 0×10^{18} and $0.5 \times 10^{18} (\Omega\ s)^{-1}$. With increase of temperature, that is, at 800 K, there was a substantial rise in σ^{ave} (T). The maximum value for $Mn_xZn_{1-x}Te$ (x = 8%) is $4.8 \times 10^{18} (\Omega\ ms)^{-1}$ and $0.4 \times 10^{18} (\Omega\ ms)^{-1}$ for up and down case and for $Mn_xZn_{1-x}Te$ (x = 16%) was $5.6 \times 10^{18} (\Omega\ ms)^{-1}$ and $0 \times 10^{18} (\Omega\ ms)^{-1}$ for up and down, respectively. At the same temperature (800 K) the compound having doping 16% showed the maximum value of electrical conductivity.

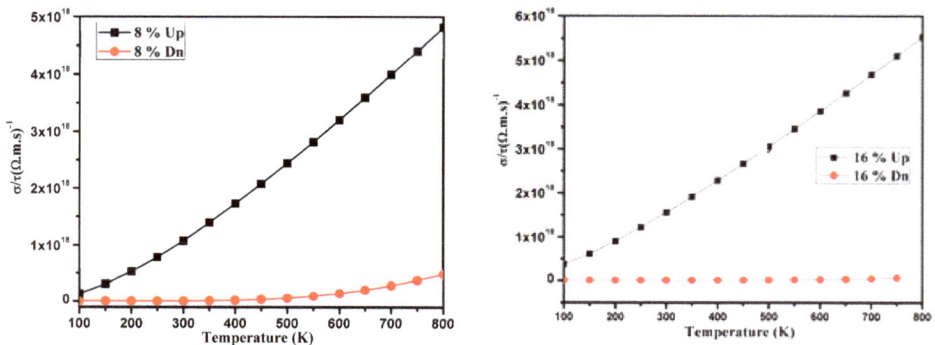

Figure 11. Electrical conductivity (long solid curve-black color for up) and (long solid curve-red color for down) for $Mn_xZn_{1-x}Te$ (x = 8% and 16%) versus temperature.

The calculated value of the Seebeck coefficient S^{ave} (T) is plotted as a function of T (100–800 K) as shown in Figure 12. The Seebeck coefficient has an inverse relation with carrier concentration, which is represented by the following formula

$$S = \left(8\pi^2 K_B/3eh^2\right) m * T \left(\pi/3n\right)^{2/3} \tag{9}$$

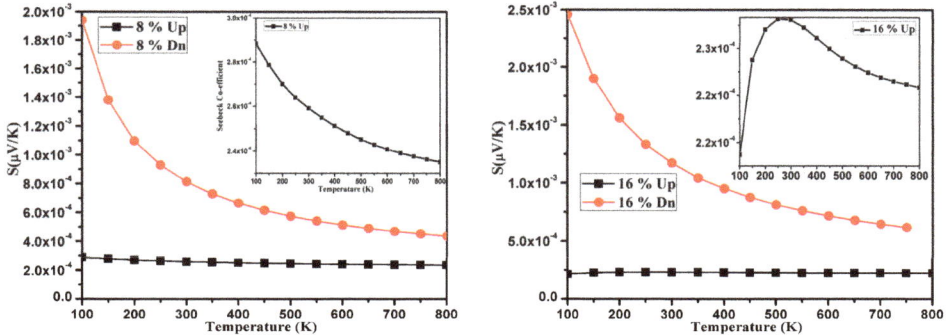

Figure 12. Temperature dependences of Seebeck coefficient (long solid curve-black color for up) and (long solid curve-red color for down) for $Mn_xZn_{1-x}Te$ (x = 8% and 16%).

In the equation above, n is carrier concentration while m* means effective mass, K_B is Boltzmann constant, h is Planks constant. The Seebeck coefficient of $Mn_xZn_{1-x}Te$ (x = 8% and 16%) compound shows a significant anisotropy in the all-inclusive within 100–800 K temperature range. These changes in S^{ave} (T) are caused by the band anisotropies of the electronic band structure. As the value of the Seebeck coefficient was positive, it demonstrated p-type semiconductor features (see Figure 12). With the increase of T from 100–800 K, the Seebeck coefficient of $Mn_xZn_{1-x}Te$ (x = 8%) decreased from 1.9×10^{-3}–8×10^{-4} μVK^{-1} up case and for down case there was no change in the Seebeck coefficient as in the variation in the temperature. For $Mn_xZn_{1-x}Te$ (x = 16%) S^{ave} (T) for 100–350 K, they had an inverse relationship with temperature.

The calculated average electronic thermal conductivity σ^{ave} (T) (the inset figure shows the zoom up state) is shown in Figure 13 for $Mn_xZn_{1-x}Te$ (x = 8% and 16%). It is clear from Figure 6 that σ^{ave} (T) for these three chalcogenides increased with T and demonstratied a huge anisotropy within the entire temperature range from 100 K to 800 K (the inset figure shows the zoom 16% down state). Figure 6 confirms that the crystal $Mn_xZn_{1-x}Te$ (x = 8%) showed minimal changes of σ^{ave} (T) for down 8% with respect to Up. σ^{ave} (T) was enhanced with increasing temperature from 100–800 K and achieved its maximal value, that is, 2.8×10^{14} W/mKs for $Mn_xZn_{1-x}Te$ (x = 8% Up) and 0.5×10^{14} W/mKs $Mn_xZn_{1-x}Te$ (x = 8% Dn) when the rate of doping was increased up to 16%. Then the value of thermal conductivity was increased in the up case, as shown Figure 13. For $Mn_xZn_{1-x}Te$ (x = 16%) the σ^{ave} (T) displays the highest value at 800 K with respect to the $Mn_xZn_{1-x}Te$ (x = 8%) 3×10^{14} W/mKs (Up) and 0.25×10^{14} W/mKs (Dn).

Figure 14 presents the calculated average power factor $S^2\sigma^{ave}$ (T) for all materials (the inset figure shows the zoom 16% down state). It was observed that the PF increases linearly for $Mn_xZn_{1-x}Te$ (x = 8% Up) with temperature, while for the down case PF only slowly varied with temperature. At 100 K, $Mn_xZn_{1-x}Te$ (x = 8%) had a magnitude 0.27×10^{11} W/mK²s (Up) and 0.0×10^{11} W/mK²s (Dn) case, when temperature increases to 800 K. Power factors in case of up and down have magnitudes 2.6×10^{11} W/mK²s and 5.5×10^{10} W/mK²s, respectively. For $Mn_xZn_{1-x}Te$ (x = 16%), power factor increased for the up case with increase in temperature, as shown in Figure 14. Hence, this might reflect the fact that this crystal may be promising for cooling devices and $Mn_xZn_{1-x}Te$ (x = 8% and 16%) shows higher greater value of $S^2\sigma^{ave}$ (T) at higher temperatures only for up case.

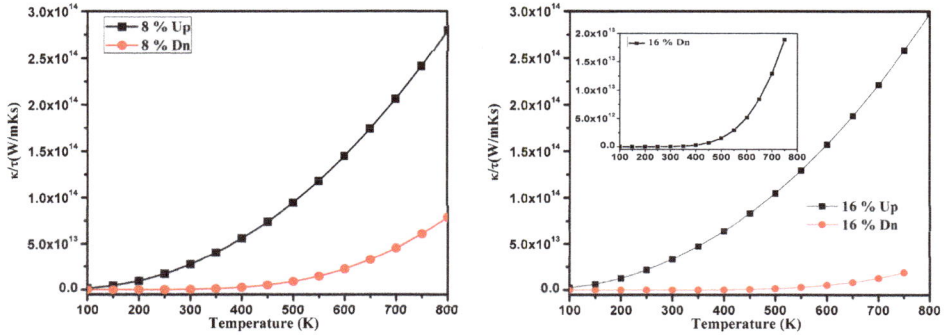

Figure 13. Thermal conductivity (long solid curve-black color for up) and (long solid curve-red color for down) dependences versus temperature for $Mn_xZn_{1-x}Te$ (x = 8% and 16 %).

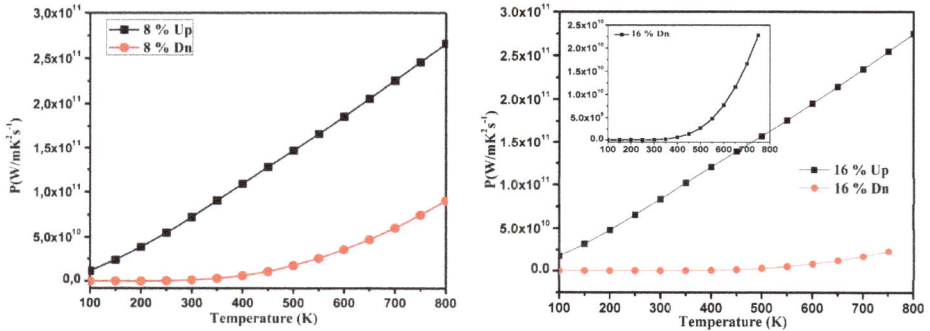

Figure 14. Temperature dependences of power factors (long solid curve-black color for up) and (long solid curve-red color for down) for $Mn_xZn_{1-x}Te$ (x = 8% and 16 %).

The figure of merit ZT = $S^2\sigma T/k$ has been calculated by including the electrical conductivity and Seebeck coefficient times T over thermal conductivity, as shown in Figure 15. For temperature range within 100 ... 800 K, ZT for $Mn_xZn_{1-x}Te$ (x = 8% and 16%) compounds exhibit different behavior, firstly both materials demonstrate a decrease with increasing T up to 800 K and after we had almost temperature-independent behavior. From the results obtained the figure of merit ZT = 1 (Up 8%) and 0.80 (Dn 8%) and for 16% its values were 0.90 and 0.60 for Up and down cases, respectively. The larger value of ZT generally is caused by lower thermal conductivity and higher electrical conductivity. In general, our calculations confirm that at higher temperatures $Mn_xZn_{1-x}Te$ (x = 8% and 16%) possess better thermoelectric efficiencies and have more potential for thermoelectric devices.

It is necessary to emphasize that a huge role for the chalcogenides begins to be played by the anharmonic phonons, as was shown using photo-induced optical methods [29–31].

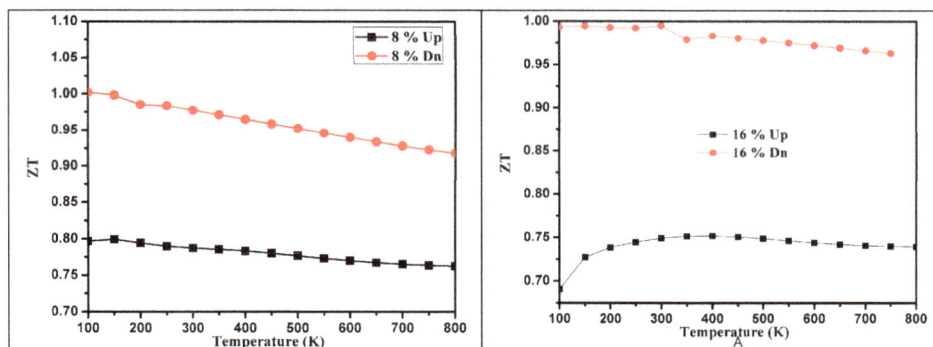

Figure 15. Temperature dependence Figure of merit (long solid curve-black color for up) and (long solid curve-red color for down) for $Mn_xZn_{1-x}Te$ (x = 8% and 16%).

4. Conclusions

The optoelectronic and transport features of the $Mn_xZn_{1-x}Te$ (x = 8% and 16%) crystals have been simulated within the DFT approach using the GGA+U approximation. It was shown that the materials studied are semiconductors with a direct type band energy gap (Γ-Γ) with magnitudes equal to 2.20 eV and 2.0 eV for $Mn_xZn_{1-x}Te$ (x = 8% and 16%), respectively. The calculated dispersions of optical functions within the energy range 0 ... 25 eV show that the up states of 8 and 16% possess strong optical response in the energy range covering the visible light and extreme UV regions. Thus, the direct band gap and strong absorption in this region of energy make these materials excellent candidates for optoelectronic devices. From the transport kinetics calculations, we have shown also that both the $Mn_xZn_{1-x}Te$ (x = 8% and 16%) crystals show promising thermoelectric properties. Finally, we expect that the current study can give several supportive hints for additional experimental investigations for the materials considered.

Author Contributions: Conceptualization, S.A.; methodology, W.K.; software, W.K.; investigation, M.I.; data curation, W.K. and S.A.; writing—original draft preparation, I.U. and M.I.; writing—review A.Y. and M.R.; and editing, I.V.K. and P.C.; supervision, I.V.K.; project administration, W.K.; funding acquisition, W.K.

Funding: This research received no external funding.

Acknowledgments: The work was supported by Project CEDAMNF, Reg. No. CZ.02.1.01/0.0/0.0/15_003/0000358, co-funded by the ERDF, GACR (Proj. 17-1484OS) and EU-COST action MP1306 (EUspec).

Conflicts of Interest: The authors declare no conflict of interest.

References

1. Mingo, N. Thermoelectric figure of merit of II-VI semiconductor nanowires. *Appl. Phys. Lett.* **2004**, *85*, 5986. [CrossRef]
2. Ghosh, B.; Ghosh, D.; Hussain, S.; Bhar, R.; Pal, A.K. Growth of ZnTe films by pulsed laser deposition technique. *J. Alloys Compd.* **2012**, *541*, 104–110. [CrossRef]
3. Ko, H.; Park, S.; An, S.; Lee, C. Intense near-infrared emission from undoped ZnTe nanostructures synthesized by thermal evaporation. *J. Alloys Compd.* **2013**, *580*, 316–320. [CrossRef]
4. Yang, X.H. Enhancing thermoelectric properties of semiconductors by heavily doping isoelectronic elements with electronegativities distinct from the host atoms. *J. Alloys Compd.* **2014**, *594*, 70–75. [CrossRef]
5. Liu, Y.; Liu, B.G. Ferromagnetism in transition-metal-doped II-VI compounds. *J. Magn. Magn. Mater.* **2006**, *307*, 245–249. [CrossRef]
6. Guo, M.; Gao, G.; Hu, Y. Magnetism and electronic structure of Mn- and V-doped zinc blende ZnTe from first-principles calculations. *J. Magn. Magn. Mater.* **2011**, *323*, 122–126. [CrossRef]

7. Saito, H.; Zayets, V.; Yamagata, S.; Ando, K. Room-Temperature Ferromagnetism in a II-VI Diluted Magnetic Semiconductor $ZN_{1-}Cr_xTe$. *Phys. Rev. Lett.* **2003**, *90*, 207202. [CrossRef]
8. Reddy, D.R.; Reddy, B.K. Laser-like mechanoluminescence in ZnMeTe-diluted magnetic semiconductor. *Appl. Phys. Lett.* **2002**, *81*, 460–462. [CrossRef]
9. Imamura, M.; Okada, A. Magnetooptical Properties of ZnMeTe Films Grown on Sapphire Substrates. *IEEE Trans. Magn.* **2006**, *42*, 3078–3080. [CrossRef]
10. Kulatov, E.; Uspenskii, Y.; Mariette, H.; Cilbert, J.; Ferrand, D.; Nakayama, H.; Ohta, H. Ab Initio study of Magnetism in III-V- and II-VI-Based Diluted Magnetic Semiconductors. *J. Supercond. Nov. Magn.* **2003**, *16*, 123–126. [CrossRef]
11. Cheng, J.; Li, D.; Cheng, T.; Ren, B.; Wang, G.; Li, J. Aqueous synthesis of high-fluorescence ZdZnTe alloyed quantum dots. *J. Alloys Compd.* **2014**, *589*, 539–544. [CrossRef]
12. Sandratskii, L.M.; Bruno, P. Density functional theory of high-T_C ferromagnetism of (ZnCr)Te. *J. Phys. Condens. Matter.* **2003**, *15*, 585–590. [CrossRef]
13. Wang, X.L.; Dou, S.X.; Zhang, C. Zero-gap materials for future spintronics, electronics and optics. *NPG Asia Mater.* **2010**, *2*, 31–38. [CrossRef]
14. Shan, C.X.; Fan, X.W.; Zhang, J.Y.; Zhang, Z.Z.; Wang, X.H.; Ma, J.G.; Lu, Y.M.; Liu, Y.C.; Shen, D.Z.; Kong, X.G.; et al. Structural and luminescent properties of ZnTe film grown on silicon by metalorganic chemical vapor deposition. *J. Vac. Sci. Technol. A* **2002**, *20*, 1886–1890. [CrossRef]
15. Rao, G.K.; Bangera, K.V.; Shivakumar, G.K. Studies on the photoconductivity of vacuum deposited ZnTe thin films. *Mater. Res. Bull.* **2010**, *45*, 1357–1360. [CrossRef]
16. Raju, K.N.; Vijayalakshmi, R.P.; Venugopal, R.; Reddy, D.R.; Reddy, B.K. Effect of substrate temperature on the structural, optical and electrical properties of vacuum-evaporated ZnTe films. *Mater. Lett.* **1992**, *13*, 336–341. [CrossRef]
17. Su, C.H.; Volz, M.P.; Gillies, D.C.; Szofran, F.R.; Lehoczky, S.L.; Dudley, M.; Yao, G.D.; Zhou, W. Growth of ZnTe by physical vapor transport and traveling heater method. *J. Cryst. Growth* **1993**, *128*, 627–632. [CrossRef]
18. Khan, M.R.H. Interface properties of a CdTe-ZnTe heterojunction. *J. Phys. D Appl. Phys.* **1994**, *27*, 2190–2193. [CrossRef]
19. Tao, I.W.; Jurkovic, M.; Wang, W.I. Doping of ZnTe by molecular beam epitaxy. *Appl. Phys. Lett.* **1994**, *64*, 1848. [CrossRef]
20. Maiti, B.; Gupta, P.; Chaudhuri, S.; Pal, A.K. Grain boundary effect in polycrystalline ZnTe films. *Thin Solid Films* **1994**, *239*, 104–111. [CrossRef]
21. Wolf, K.; Stanzl, H.; Naumov, A.; Wagner, H.P.; Kuhn, W.; Hahn, B.; Gebhardt, W. Growth and doping of ZnTe and ZnSe epilayers with metalorganic vapour phase epitaxy. *J. Cryst. Growth* **1994**, *138*, 412–417. [CrossRef]
22. Neumann-Spallart, M.; Konigstein, C. Electrodeposition of zinc telluride. *Thin Solid Films* **1995**, *265*, 33–39. [CrossRef]
23. Bozzini, B.; Lenardi, C.; Lovergine, N. Electrodeposition ofstoichiometric polycrystalline ZnTe on n^+-GaAs and Ni-P. *Mater. Chem. Phys.* **2000**, *66*, 219–228. [CrossRef]
24. Jun, Y.; Kim, K.J.; Kim, D. Electrochemical synthesis of cu-doped znte films as back contacts to cdte solar cells. *Met. Mater.* **1999**, *5*, 279–285. [CrossRef]
25. Arico, A.S.; Silvestro, D.; Antonucci, P.L.; Giordano, N.; Antonucci, V. Electrodeposited Thin Film ZnTe Semiconductors for Photovoltaic Applications. *Adv. Perform. Mater.* **1997**, *4*, 115–125. [CrossRef]
26. Karazhanov, S.Z.; Ravindran, P.; Kjekshus, A.; Fjellvag, H.; Svensson, B.G. Electronic structure and optical properties of Zn X (X = O, S, Se, Te): A density functional study. *Phys. Rev. B* **2007**, *75*, 155104. [CrossRef]
27. Huang, M.Z.; Ching, W.Y. Calculation of optical excitations in cube semiconductors. I. Electronic structure and linear response. *Phys. Rev. B* **1993**, *47*, 9449. [CrossRef]
28. Subhan, F.; Azam, S.; Khan, G.; Irfan, M.; Muhammad, S.; Al-Sehemi, A.G.; Naqib, S.H.; Khenata, R.; Khan, S.; Kityk, I.V.; et al. Elastic and optoelectronic properties of $CaTa_2O_6$ compounds: Cubic and orthorhombic phases. *J. Alloys Compd.* **2019**, *785*, 232–239. [CrossRef]
29. Kityk, I.V. IR-induced Second Harmonic Generation in Sb_2Te_3-BaF_2-$PbCl_2$ Glasses. *J. Phys. Chem. B* **2003**, *107*, 10083–10087. [CrossRef]

Crystals **2019**, *9*, 247

30. Kityk, I.V. IR-stimulated second harmonic generation in Sb_2Te_2Se-BaF_2-$PbCl_2$ glasses. *J. Mod. Opt.* **2004**, *51*, 1179–1189.

31. Shpotyuk, O.I.; Kityk, I.V.; Kasperczyk, J. Mechanism of reversible photoinduced optical effects in amorphous As_2S_3. *J. Non Cryst. Solids* **1997**, *215*, 218–225. [CrossRef]

crystals

MDPI

Article

Van der Waals Density Functional Theory vdW-DFq for Semihard Materials

Qing Peng [1,2,3,*], **Guangyu Wang** [4], **Gui-Rong Liu** [4] and **Suvranu De** [1]

[1] Department of Mechanical, Aerospace and Nuclear Engineering, Rensselaer Polytechnic Institute, Troy, NY 12180, USA; des@rpi.edu

[2] School of Power and Mechanical Engineering, Wuhan University, Wuhan 430072, Hubei, China

[3] Nuclear Engineering and Radiological Sciences, University of Michigan, Ann Arbor, MI 48109, USA

[4] School of Aerospace Systems, University of Cincinnati, Cincinnati, OH 45221, USA; wanguc2011@hotmail.com (G.W.); liugr@uc.edu (G.-R.L.)

* Correspondence: qpeng.org@gmail.com

Received: 9 April 2019; Accepted: 2 May 2019; Published: 8 May 2019

Abstract: There are a large number of materials with mild stiffness, which are not as soft as tissues and not as strong as metals. These semihard materials include energetic materials, molecular crystals, layered materials, and van der Waals crystals. The integrity and mechanical stability are mainly determined by the interactions between instantaneously induced dipoles, the so called London dispersion force or van der Waals force. It is challenging to accurately model the structural and mechanical properties of these semihard materials in the frame of density functional theory where the non-local correlation functionals are not well known. Here, we propose a van der Waals density functional named *vdW-DFq* to accurately model the density and geometry of semihard materials. Using β-cyclotetramethylene tetranitramine as a prototype, we adjust the enhancement factor of the exchange energy functional with generalized gradient approximations. We find this method to be simple and robust over a wide tuning range when calibrating the functional on-demand with experimental data. With a calibrated value $q = 1.05$, the proposed vdW-DFq method shows good performance in predicting the geometries of 11 common energetic material molecular crystals and three typical layered van der Waals crystals. This success could be attributed to the similar electronic charge density gradients, suggesting a wide use in modeling semihard materials. This method could be useful in developing non-empirical density functional theories for semihard and soft materials.

Keywords: density functional theory; van der Waals corrections; semihard materials; molecular crystals

1. Introduction

Semihard materials generally consist of molecules. Energetic materials are a common class of semihard materials with light composites of elements C, H, O, and N, of which the molecular sizes are moderate (ranging from 10 to 1000 atoms) leading to fast chemical kinetics so that the chemical reactions during decomposition or detonation occurs in a few picoseconds. To gain high energy densities, these molecules are mostly bonded via intermolecular London dispersion forces to form molecular crystals (Figure 1a–k), which belong to semihard materials since they have mild hardness. The mass density is closely related to the energy density and detonation performance. Therefore, the accurate prediction of the mass density within 3% of experimental values is critical in predicting the reaction rate and energy density. In general, Density Functional Theory (DFT) [1] is a reliable and popular tool to predict material properties and material designs. However, the London dispersion force is poorly described with conventional semi-local approximations to DFT [2] since it is essentially a nonlocal electron correlation effect. As a result, an accurate description of the dispersion force is essential in DFT modeling of energetic materials.

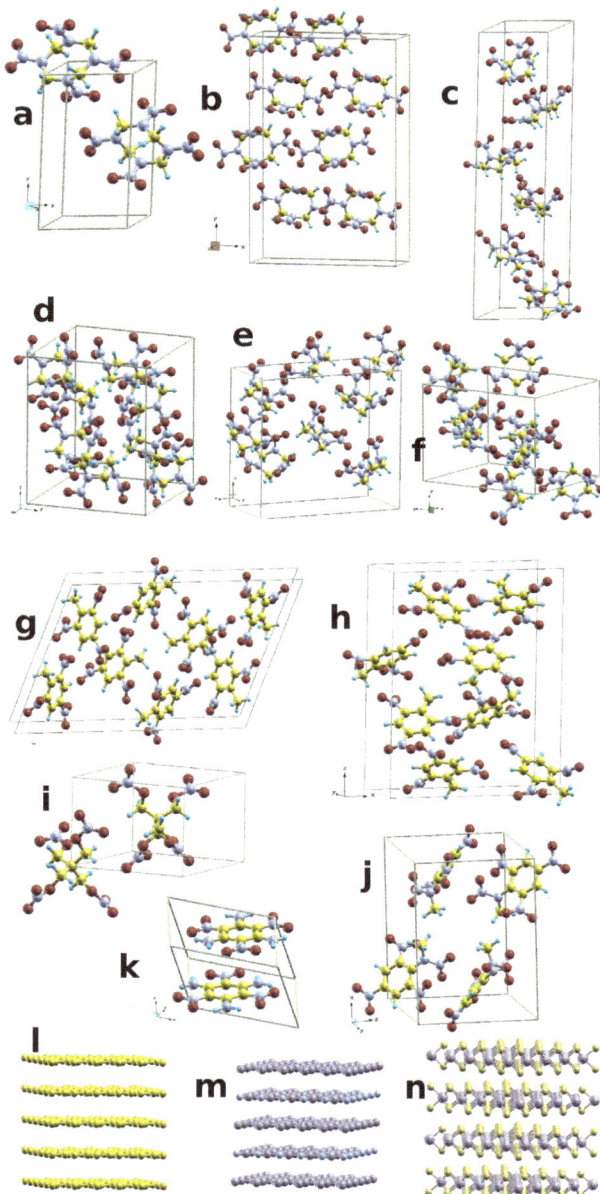

Figure 1. Semihard materials. The primitive unit cell of (**a**) β-HMX, (**b**) α-HMX, (**c**) δ-HMX, (**d**) α-RDX, (**e**) β-RDX, (**f**) γ-RDX, (**g**) α-TNT, (**h**) β-TNT, (**i**) PETN, (**j**) Tetryl, and (**k**) TATB. The layer materials are (**l**) Graphite, (**m**) h-BN, and (**n**) MoS_2 .

There are extensive studies to improve the modeling of dispersion, or van der Waals (vdW) interactions [2,3], such as DFT-D [4–9], vdW-TS [10,11], DFT-vdW^{surf} [12], and vdW-DF [13–16] methods. The DFT-D1 and DFT-D2 methods are classified as the 1st rung of Jacob's Ladder for DFT-based dispersion correction schemes according to Klimes and Michaelides [3]. The DFT-D3 and vdW-TS methods belong to the next rung that has environment dependent C_6 corrections. Non-local density functionals like DFT-DF and DFT-DF2 methods sit in the 3rd rung [3], of which the overall accuracy increases, as well as the computational demand. Recently, a few improvements in vdW-DF methods have been proposed, including optPBE-vdW, optB88-vdW, optB86b-vdW [17,18], vdW-DF2-C09 [19], vdW-DF-cx [20], rev-vdW-DF2 [21], PBE+rVV10 [22], PBEsol+rVV10 [23], SCAN+rVV10 [24], and SG4+rVV10 [25]. These methods are designed to minimize the error of the binding energies and interaction energy curves in a certain collection of materials, such as the S22 data set [26], aiming for a general use for all materials. The accuracy of these models on energetic materials like HMX has not been assessed yet. We found that the prediction of the mass density in HMX is still not satisfied, as opposed to the claimed success in many other materials. Here we propose a vdW-DF-based method (we denote it as vdW-DFq) using a single parameter to tune the exchange functional so that the target mass density can be achieved. We are looking for a reliable model for a particular material, instead of a general formula for all materials.

The cyclotetramethylene tetranitramine (HMX) shown in Figure 1a–c is an important secondary explosive. The P2$_1$/c monoclinic crystal structure, namely β-phase, (Figure 1a) is thermodynamically the most stable phase of HMX under ambient conditions. β-HMX has a high detonation velocity and high energy density. HMX is most commonly used in military and industrial applications including automobile air bags, rocket propellants, and polymer-bonded explosives. Therefore, β-HMX has been extensively studied [27–29]. We focus on β-HMX as a representative of semihard materials because of its moderate bulk modulus (around 15 GPa).

2. Methodology

Within the DFT frame work, the van der Waals density functional (vdW-DF) [13] is good for both covalent and van der Waals (vdW) interactions in a seamless fashion. The exchange correlation energy E_{xc} as sum of exchange energy E_x and correlation energy E_c is expressed as

$$E_{xc} = E_x + E_c = E_x^{GGA} + E_c^{LDA} + E_c^{NL} \tag{1}$$

where E_x^{GGA} is the exchange energy within the generalized gradient approximation (GGA) [30], E_c^{LDA} is the short-range correlation energy within the local density approximation (LDA) [31], and E_c^{NL} is the non-local correlation energy is long-range beyond LDA. E_c^{NL} could be expressed as

$$E_c^{NL} = \frac{1}{2} \int \int d\mathbf{r} d\mathbf{r}' n(\mathbf{r}) \phi(d, d') n(\mathbf{r}'), \tag{2}$$

where $n(\mathbf{r})$ is the electron density at the point $\mathbf{r}$, and $\phi(d, d')$ is the vdW kernel [13], $d = q_0(\mathbf{r})|r - r'|)$, $d = q_0(\mathbf{r}')|r - r'|)$, q_0 is a function of $n(\mathbf{r})$ and its gradient $|\nabla n(\mathbf{r})|$, which determines the long-range asymptote as well as the short-range damping of the vdW kernel. The choice of LDA for the short-range correlation is to avoid possible double counting of the gradient on $n(\mathbf{r})$ and $|\nabla n(\mathbf{r})|$. Ultimately, this choice implicitly implies that the only non-local correlation term is due to vdW interactions, where the vdW kernel is actually derived in a way to model dispersion. The computational cost for vdW-DF calculations is marginal as compared with orbital dependent functionals and recent developments in efficient algorithms [32–34]. As a result, vdW-DF calculations are feasible in large calculations at a computational cost comparable with that of conventional semi-local DFT, at about only 30% additional computation time from the authors' experience.

The general form of the GGA exchange energy functional is given as

$$E_x^{GGA} = \int d\mathbf{r} n(\mathbf{r}) \epsilon_x^{unif}[n(\mathbf{r})] F_X^{GGA}[s(\mathbf{r})], \tag{3}$$

where $\epsilon_x^{unif}(n) = -3k_F/4\pi$ with $k_F = (3\pi^2 n)^{\frac{1}{3}}$ is the exchange energy for the uniform electron gas, $F_X^{GGA}(s)$ is the GGA enhancement factor, which distinguishes one GGA from another. The variable s in $F_X^{GGA}(s)$ is the dimensionless reduced gradient, defined as

$$s = \frac{|\nabla n|}{2(3\pi^2 n)^{1/3} n}. \tag{4}$$

The reduced gradient expresses how fast the density varies on the scale of the local Fermi wavelength k_F. When s is set to zero, the GGA functional reduces to LDA. There are quite a few proposals of the E_x^{GGA} for vdW-DF [17–19,32].

The nonlocal vdW-DF consistently overestimates the inter-fragment distance, although it improves the description of van der Waals interactions over conventional semi-local DFT. The original vdW-DF uses the revised Perdew-Burke-Ernzerhof (revPBE) [35] exchange functional for weakly interacting molecular pairs. However, the revPBE exchange correction is too repulsive at small separations [36]. Therefore, a few efforts are devoted to develop a better exchange functional for the nonlocal correlation functional. Klimeš et al. [17,18] have introduced several exchange functionals with accurate binding energies of molecular duplexes in the S22 data set [37]. Cooper constructed an exchange functional (named C09x) based on revPBE [19]. The second version of vdW-DF (vdW-DF2) uses the revised PW86 (rPW86) exchange functional [14]. Additionally, there are a few other nonlocal correlation functionals [38–41].

Recently a vdW-DF optimized semi-empirically based on the Bayesian error estimation named BEEF was proposed [42]. Hamada also proposed an exchange functional for the second version of vdW-DF (vdW-DF2). The so called rev-vdW-DF2 [21] is based on the GGA exchange functional of optB86b [18] with a revised q value of 0.7114 (as named B86R). A very recent study suggests that rev-vdW-DF2 is very accurate for weakly bound solids overall compared to other methods. However, it is still not satisfactory on our demanded accurancy in geometry. Nevertheless, the performance of the improved functionals depend on the system, and for the applications to a variety of systems including surfaces and interfaces, which consist of materials with different bonding natures, accuracy of vdW-DF has yet to be improved.

It is worth noting that all these methods are targeting general materials and all the properties, which is definitely a non-trivial work because the exchange enhancement factor is hard to satisfy in all systems where the reduced density gradient is different from case to case. In some cases, one needs a well tailored density functional only for a particular property of interest. For example, the mass density is the primary concern in prediction of the detonation performance of energetic materials. Thus, a relatively simple, accurate, and tunable density functional theory is indispensable.

To obtain an accurate exchange functional for vdW-DF, the following properties are considered here. First, E_x^{GGA} should recover the second-order gradient expansion approximation (GEA) [43] at the slowly varying density limit as

$$F_X^{GEA}(s) = 1 + \mu s^2, \tag{5}$$

with $\mu = 10/81$. This property plays an important role in predicting equilibrium geometries for solids and surfaces [44]. This constraint has been applied in other exchange functionals [18,20,45]. It is worth pointing out that C09 employs $\mu = 0.0864$. Second, at large gradient limit, it proposed that F_x should have $s^{2/5}$ dependence so that it can avoid the spurious binding from exchange only. So far only Perdew and Wang (PW86) [46] and Becke (B86b) [47] exchange functionals fulfill this reqirement [45].

Murray et al. [45] re-parameterized PW86 to satisfy these known limits (PW86R), which leads to vdW-DF2. However, it turns out that the exchange functional is repulsive in solids and adsorption systems, leading to overestimated equilibrium distances [48].

Another efforts is that Hamada proposed a revised B86b exchange functional (B86R), which improves the description of the attractive van der Waals interactions near the equilibrium over the original vdW-DF. The enhancement factor for the B86b exchange functional is given by

$$F_X(s) = 1 + f_q \mu_0 s^2, \tag{6}$$
$$f_q = \frac{1}{(1+\mu_0 s^2/q)^{4/5}} \tag{7}$$

When $s \to 0$, the factor $f_q = 1$ and the first condition is fulfilled. The value of q is arbitrarily determined by a least square fit to F_X of the original B86b at $8 < s < 10$.

The first order derivative of the exchange enhancement factor with respect to the reduced density gradient $\frac{d}{ds}F_X$ is important for binding separations [20], which can be obtained directly, as

$$\frac{d}{ds}F_X(s) = 2\mu_0 f_q s - 1.8\mu_0^2 f_q^{9/4}/qs^3. \tag{8}$$

We can determine q by reproducing the B86b asymptote at $s \to \infty$. However, the results are still questionable in improving accuracy. Thus Hamada have obtained $q = 0.7114$ by the least square fit. Using the same principle, Berland and Hyldgaard [20] introduce an exchange functional, namely LV-PW86r, for the original vdW-DF which leads to vdW-DF-cx.

Here we propose the vdW-DFq method to tune the q for the particular system as a calibration to the experiments. For instance, we find a very good agreement with experiment in predicting the mass density of β-HMX using $q = 1.05$. Figure 2a,b illustrates the enhancement factor of B86q with the $q = 1.05$ exchange energy functional along with ten other corresponding ones in the low-gradient (panel a) and high-gradient domains (panel b). Their derivatives are displayed in the lower part of the figure.

It is desirable that the exchange energy functional fullfills the gradient at the slowly varying density limit, in addition to the revPBE asymptote at the large reduced gradient. Therefore, we roughly divide the range of the reduced density gradient into two domains: 0–3 for the slowly varying density domain (low-gradient domain) and 3–10 for the large reduced gradient domain (high-gradient domain). Plots of the enhancement factors provide a way to visualize the s dependence of the GGA.

A universal accurate function of the enhancement factor might be hard find. However, it is practical to tune the enhancement factor for a particular kind of materials that have similar electronic charge density gradients s. To achieve that, a function of enhancement factor that can be tuned in a large range is expected to explore the more $F_X - s$ and $\frac{d}{ds}F_X - s$ space. We find that the enhancement factor in form of Equation (5) are surprisingly good for tuning enhancement factors by varying q values. One can see in Figure 3 that both the enhancement factor and its derivative vary over a wide range when q is changed from 0.1 to 100.0. This illustrates that q is an effective variable to tune the exchange energy functional and binding separations.

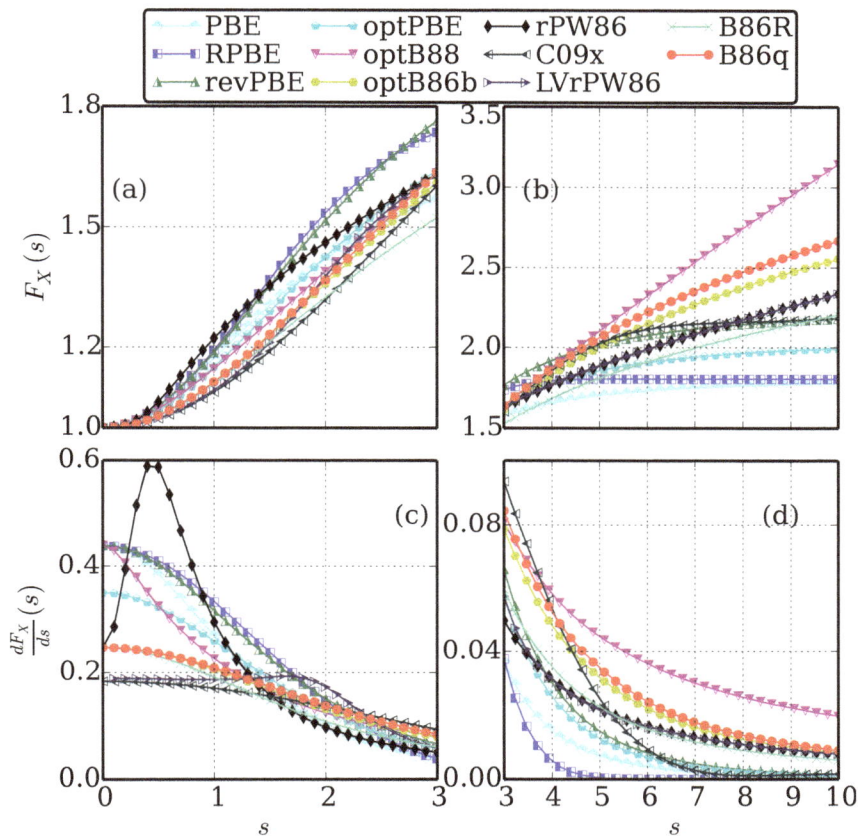

Figure 2. Enhancement factors. The exchange enhancement factor F_X as a function of reduced density gradient s for 11 exchange functionals in GGA formula in the range of (**a**) 0–3 for low-gradient domain and (**b**) 3–10 for high-gradient domain. The vdW-DFq used the optB86b exchange energy with $q = 1.05$. The derivatives $\frac{d}{ds} F_X$ as a function of s are illustrated in the range of (**c**) 0–3 and (**d**) 3–10.

3. Results and Analysis

The proposed vdW-DFq method is firstly tested for β-HMX, whose primitive cell is shown in Figure 1a. The geometry optimization of the primitive unit cell starts with the experimental configuration [49]. The atoms are then relaxed to the configuration with the minimum total energy. Our optimized crystal structure has a space group of $P2_1/c$. The optimized lattice constants using various methods are summarized in Table 1. The relative standard deviation (RSD) or the error of the predicted volumes to the experiment are also listed as a measurement of the accuracy of these methods.

The experimental values are taken from the report of Herrmann et al. [50] in which they measured the crystal structures under various temperatures using X-ray diffraction. The standard DFT calculations (denoted as PBE hereafter) without van der Waals corrections agree well with the previous theoretical prediction [27]. It shows that standard DFT calculations give poor predictions. For example, the volume of the unit cell is 6.8% larger than the experimental value. The volume of β-HMX from rung 1 vdW methods (PBE-D2), rung 2 vdW methods (PBE-D3, PBE-TS, RPBE-D3), and PBEsol calculations [51] has a significant improvement over standard DFT calculations. They have a RSD within 2% compared to the experimental value.

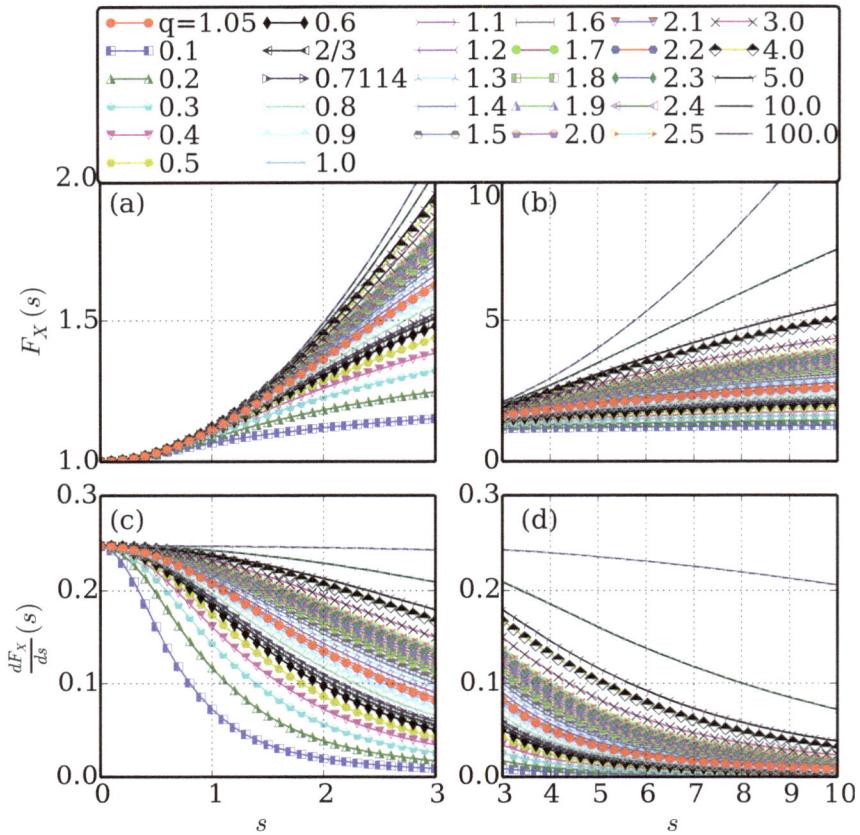

Figure 3. q dependent enhancement factor. The comparison of the exchange enhancement factor $F_X(s)$ in exchange energy functionals in the vdW-DFq method for various q values. The reduced density gradient s is in the range of (**a**) 0–3 low-gradient domain and (**b**) 3–10 high-gradient domain. The B86q and B86R denotes the case of $q = 1.05$ and 0.7114 respectively. The $\frac{d}{ds} F_X$ is illustrated in the range of (**c**) 0–3 and (**d**) 3–10.

The van der Waals density functional methods are in rung 3. The explicitly tested methods are original vdW-DF [13], vdW-DF2 [14], optPBE-vdW [18], optB88-vdW [18], optB86b-vdW [18], vdW-DF2-C09 [19], LV-PW86r, vdW-DF-cx [20], rev-vdW-DF2 [18], screened-vdW [52], and vdW-DFq. It is worth pointing out that the gradient coefficient Z_{ab} is -0.8491 and -1.887 in vdW-DF and vdW-DF2 methods, respectively [18]. One might expect that the more computing intensive vdW-DF models including the latest rev-vdW-DF2 method have higher accuracy as rung 3 vdW methods compared to rung 0, 1 and 2 methods. However, it turns out that the accuracy some of the vdW-DF functionals are even worse than standard PBE calculations as shown in Table 1. This implies that the exchange energy functional plays an important role in determining the geometry of semihard materials in vdW-DF methods. Only a carefully selected exchange energy functional together with the vdW-DF methods can predict the volume accurately.

It is also seen from the results listed in Table 1 that the predicted volume scatters over a wide range. One might want a way to monotonically tune the volume to that of the experimental volume. Using the one-parameter vdW-DFq method proposed here, one can smoothly tune the system value to the target one. It is worth noting that the only difference between the rev-vdW-DF2 and the vdW-DFq methods is the q value, which is fixed as 0.7114 in rev-vdW-DF2, but tunable in vdW-DFq from system to system, where a value of $q = 1.05$ is good for β-HMX in this work, as illustrated in Figure 4. In addition, we have carried out the calculation using optB86q-GGA without vdW-DF2 corrections to examine the effect of vdW-DF2 corrections to the system's volume. We find that the volume will be 19.75% larger than the experiment, clearly showing that vdW-DF2 corrections are important.

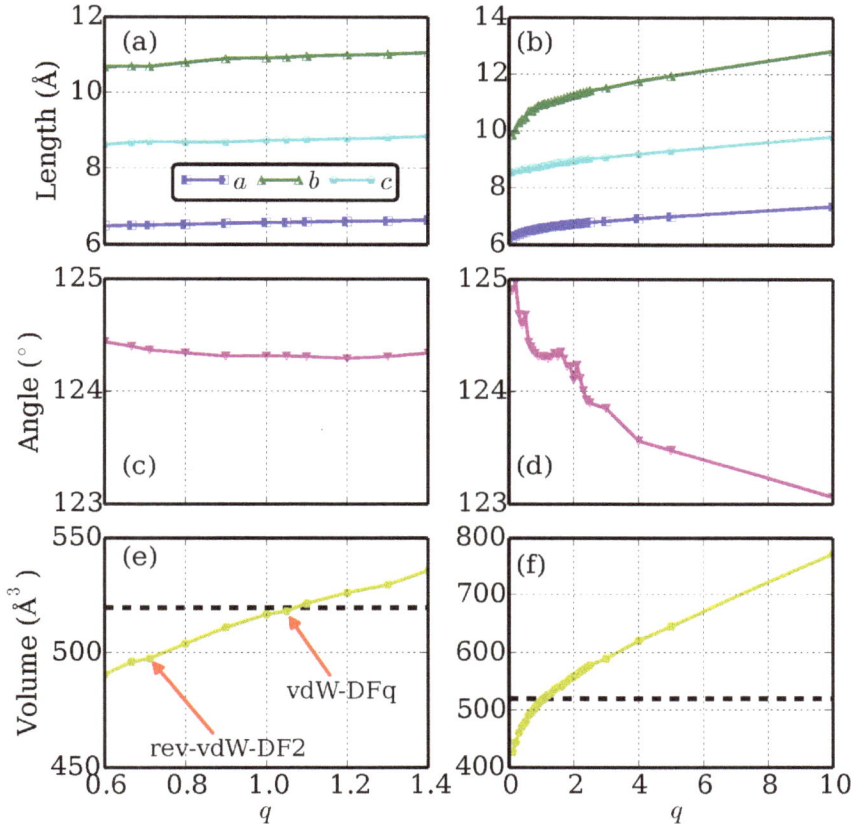

Figure 4. Geometry of β-HMX. The lattice constants a,b,c (**a,b**), lattice angle β (**c,d**), and volume of the unit cell V (**e,f**) predicted from the proposed vdW-DFq method as a function of the one-parameter q in a small range around 1.0 (**left**) and large range from 0 to 10 (**right**). The B86q and B86R denotes the case of $q = 1.05$ and 0.7114 respectively. The dashed line is the experimental volume.

It is seen from Figure 4 that the lattice constants, lattice angle, and the volume can be tuned continuously at a very large range. For example, the volume varies from 82% to 200% when q changed from 0.1 to 100.0. Within the vicinity of the $q = 1.0$, the volume varies linearly with respect to q. $q = 1.05$ gives the best match to the experimental volume with RSD of -0.16%. As a comparison, the rev-vdW-DF2 method [18], the latest vdW-DF series method, predicted the volume as 496.905 Å^3, 4.2% smaller than the experiment, which falls out of the satisfactory scope of 3%. At large deviation from 1.0, the relationship between the volume and the variable q is nonlinear as expected.

The large tunable range and linear behavior in the vicinity of the q corresponding to the experimental volume make the calibration of the vdW-DF much easier. This also indicates that our method is simple and robust in modeling various semihard materials where the vdW interactions are critical but the analytical formula is unknown *a priori*. We focus on the geometry study because the density is our main concern. One could tune q for other properties, such as for binding and cohesive energies. One needs to keep in mind that the tuning parameter q is not unique: there are quite a few parameters that affect the exchange energies. For example, one can also vary the gradient coefficient Z_{ab} to modify the exchange energy functionals and it could tune the system's volume as well. Our test shows in Table 1 that when $Z_{ab} = -0.8491$ (vdW-DFk') is used instead of -1.887 in vdW-DFq, the system volume will have a -4.9% change.

In the proposed vdW-DFq method, the $q = 1.05$ is calibrated to β-HMX with the primitive unit cell volume. A general concern rises about the predictive power of this method. To that end, we further study the other semihard materials utilizing vdW-DFq method. Additional ten common structures of high explosives are investigated as α and δ-HMX [53], α [54], β [55], and γ-RDX [56], α (Monoclinic) and β (Orthorhombic)-TNT [57], PETN [58], Tetryl [59], and TATB [60]. We focus on the mass density as it is the predominant factor for detonation density and pressure. The results of the lattice parameters, volume, mass density, and density RSD of these 10 structures are summarized in Table S1 in the Supplementary Information, compared with the predictions from vdW-DF2 and rev-vdW-DF2 methods as well as experiments. The relative standard deviations of mass densities are referred to the corresponding experiments. Particularly, we plotted the mass density RSD in Figure 5.

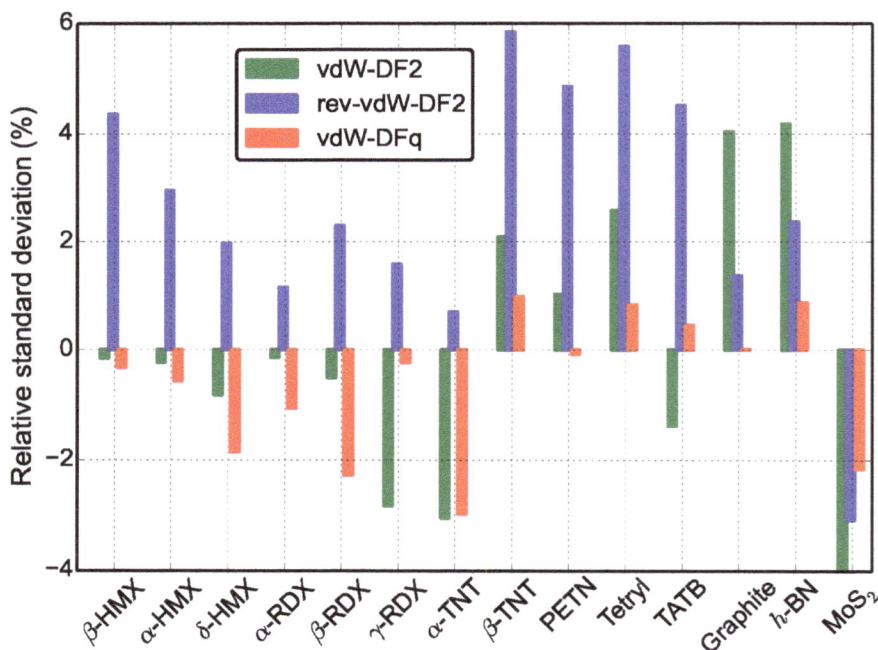

Figure 5. Relative standard derivative of the predicted mass density of the semihard materials. The mass density RSD referring to the experiments using vdW-DF2, rev-vdW-DF2, and vdW-DFq method for 11 common energetic materials molecular crystals and 3 typical layered van der Waals crystals.

It shows that the recent developed rev-vdW-DF2 method is worse than conventional vdW-DF2 method in predicting the mass density of the energetic materials, and the proposed vdW-DFq method has the best performance of the examined three vdW-DF methods. The root mean square error of densities predicted from the vdW-DF2, rev-vdW-DF2, and vdW-DFq methods for the examined 11 molecular crystals is 1.7%, 3.7%, and 1.4%, respectively. These results indicate that vdW-DFq method with $q = 1.05$ might be a good first-principles approach to study energetic materials. This success could be attributed to the similar electronic charge density gradients, because all these materials are composed of similar elements of carbon, nitrogen, oxygen, and hydrogen, and all have similar mass densities around 1.9 g/cm^3.

In addition to energetic materials, we further study the mass density of the layered van der Waals crystals [61,62]. Graphite, hexagonal boron nitride (*h*-BN), and molybdenum disulfide (MoS$_2$) are typical layered crystals. Their lattice parameters, volume, mass density, and density RSD are summarized in Table S1 in the Supplementary Information.

As shown in Figure 5, the vdW-DFq method prediction also has good agreements with experimental measurements. The performance of the vdW-DFq method is better than the vdW-DF2 and rev-vdW-DF2 methods in the layered van der Waals crystals. For the total 14 structures, the root mean square error of densities is 5.52%, 3.46%, and 1.37% for the vdW-DF2, rev-vdW-DF2, and vdW-DFq methods, respectively. The better performance of the vdW-DFq method over the vdW-DF2 and rev-vdW-DF2 methods in predicting the geometries of the 14 representative molecular crystals implies that the proposed vdW-DFq method could also have good performance in modeling other semihard materials. It is reasonable because the q value is calibrated and thus "localized" to semihard materials, as opposed to the other two methods which aim at a universal method for all materials.

It is worth mentioning that graphite, *h*-BN, and MoS$_2$ are actually highly anisotropic materials. The success of our vdW-DFq method also suggests that, in some cases, finding correct equilibrium configurations does not necessarily imply finding the correct dispersion interactions [63], but a compromise of various approximations, which, however, vary from case to case. From this aspect, it evidences the essential role of q that is tunable to optimize such an accommodation. It might be expected that vdW-DFq method is a handy tool for various vdW systems on-demand.

4. Conclusions

In summary, we assess the performance of van der Waals (vdW) density functionals in predicting the geometry of semihard materials. We propose a one-parameter empirical van der Waals density functional named vdW-DFq to continuously tune the lattice constants by adjusting the enhancement factor of the exchange energy functionals. We illustrate that the lattice constants and volumes can be tuned over a wide range, and with a linear behavior in the vicinity of the q corresponding to the experimental volume. The application on β-HMX shows that our method is simple and robust in modeling various semihard materials. By one-parameter tuning, the vdW density functional can be well calibrated to the experimental value. This method provides a controllable approach to accurately model the mechanical, electrical, chemical, and other properties within the framework of van der Waals density functionals. Further study on the 10 energetic material molecular crystals and 3 layered van der Waals crystals shows that the proposed vdW-DFq method has better performance than the vdW-DF2 and rev-vdW-DF2 methods in predicting the geometries, which indicates that the vdW-DFq method could be good at modeling semihard materials. Our investigation might also inspire computational physicists to develop a better non-empirical vdW density functional. Nevertheless, our vdW-DFq method is a on-demand handy tool for various vdW systems by tuning q value.

Table 1. Geometry. Lattice constants a, b, c, lattice angle β, and volume of the unit cell V from various methods along with the relative standard deviation (RSD%) compared with experiments.

	a(Å)	b(Å)	c(Å)	β	V(Å^3)	RSD
Expt. [a]	6.537	11.054	8.7018	124.443	518.558	
PBE [b]	6.673	11.312	8.894	124.395	553.99	6.8
PBE-D2 [b]	6.542	10.842	8.745	124.41	511.73	−1.4
PBE-D3 [c]	6.541	10.894	8.748	124.38	514.488	−0.8
PBE-TS [c]	6.530	10.985	8.770	124.572	517.992	−0.1
RPBE-D3 [c]	6.580	11.061	8.762	124.357	526.501	1.5
PBEsol [c]	6.603	10.928	8.709	124.002	520.949	0.5
vdW-DF	6.653	10.096	8.901	124.756	539.880	4.1
vdW-DF2	6.575	10.946	8.765	124.579	519.387	0.2
optB88-vdW	6.476	10.662	8.680	124.735	492.535	−5.0
optB86b-vdW	6.481	10.635	8.700	124.749	492.710	−5.0
optPBE-vdW	6.550	10.877	8.760	124.703	513.088	−1.1
vdW-DF-C09	6.435	10.494	8.650	124.807	479.676	−7.5
vdW-DF-cx	6.556	10.787	8.779	124.557	511.324	−1.4
rev-vdW-DF2 [d]	6.504	10.696	8.685	124.670	496.905	−4.2
vdW-DFq [e]	6.569	10.921	8.737	124.316	517.742	−0.16
optB86q [f]	6.922	11.773	8.985	121.986	620.991	19.8
vdW-DFk′ [g]	6.486	10.636	8.694	124.713	493.013	−4.9

[a] Ref. [50]; [b] Ref. [51]; [c] Ref. [64]; [d] $q = 0.7114$; [e] $q = 1.05$; [f] optB86q-GGA calculations only (without vdW-DF2 corrections); [g] using $Z_{ab} = -0.8491$.

5. Computational Method

We used a conventional unit cell with two HMX molecules. There are 56 atoms in total in one unit cell. The periodic boundary conditions are applied in three normal directions. The first-principles calculations based on density-functional theory (DFT) were carried out with the Vienna Ab-initio Simulation Package (VASP) [65]. The Kohn-Sham Density Functional Theory (KS-DFT) [1] is employed with the generalized gradient approximation as parameterized by Perdew, Burke, and Ernzerhof (PBE) for exchange-correlation functionals [30]. We have examined two additional exchange-correlation functionals: revised Perdew-Burke-Ernzerhof (RPBE) [66], and Perdew-Burke-Ernzerhof revised for solids (PBEsol) [44].

The electrons explicitly included in the calculations are the $1s^1$ for hydrogen atoms, $2s^2 2p^2$ for carbon atoms, $2s^2 2p^3$ for nitrogen atoms, and $2s^2 2p^4$ for oxygen atoms. The core electrons are discribed by the projector augmented wave (PAW) and pseudo-potential approach [67,68]. A energy cutoff of 520 eV is applied in this study. All the vdW-DF calculations were done self-consistently using the efficient algorithm of Román-Pérez and Soler [33], implemented in VASP by Klimeš et al. [18]. It is worth noting that it is not evident a priori that self consistency should necessarily lead to higher accuracy: although self-consistent, vdW-DF and related methods still imply a number of approximations which may be way larger than self-consistency vdW effects. One of these is the lack of long-range many body effects which may amount to more than 10% of the total dispersion energy (see [69–71], and can lead to unexpected features in highly anisotropic systems [72,73].

The criterion electronic relaxation is 10^{-6} eV. The optimized atomic geometry was achieved when the forces on each atom is smaller than 0.001 eV/Å. The irreducible Brillouin Zone was sampled with a $5 \times 3 \times 4$ Gamma-centered k-mesh. The other 13 structures are optimized using similar parameters in DFT calculations.

Supplementary Materials: The following are available at http://www.mdpi.com/2073-4352/9/5/243/s1, Supplementary information includes the lattice parameters, volume, mass density, and density relative standard deviation of other 13 structures and is available in the online version of the paper. Correspondence and requests for materials should be addressed to Q.P. (qpeng.org@gmail.com). About how to use our method in VASP. We therefore write a "howto" file, as the supplementary.

Author Contributions: Conceptualization, Q.P.; computation, Q.P., G.W.; writing—original draft preparation, Q.P.; writing—review and editing, Q.P., G.W, G.-R.L., S.D.; supervision, S.D.; project administration, S.D.; funding acquisition, S.D., G.-R.L.

Funding: The authors would like to acknowledge the generous financial support from the Defense Threat Reduction Agency (DTRA) Grant # HDTRA1-13-1-0025 and the Office of Naval Research grants ONR Award # N00014-08-1-0462 and # N00014-12-1-0527.

Acknowledgments: We thank Shengbai Zhang, Jianwei Sun, and Chen Huang for helpful discussions. Computational resources were provided by Rensselaer Polytechnic Institute through AMOS, the IBM Blue Gene/Q system at the Center for Computational Innovations.

Conflicts of Interest: The authors declare no competing interests.

References

1. Hohenberg, P.; Kohn, W. Inhomogeneous electron gas. *Phys. Rev.* **1964**, *136*, B864. [CrossRef]
2. Hermann, J.; DiStasio, R.A., Jr.; Tkatchenko, A. First-Principles Models for van der Waals Interactions in Molecules and Materials: Concepts, Theory, and Applications. *Chem. Rev.* **2017**, *117*, 4714–4758. [CrossRef]
3. Klimes, J.; Michaelides, A. Perspective: Advances and challenges in treating van der waals dispersion forces in density functional theory. *J. Chem. Phys.* **2012**, *137*, 120901. [CrossRef] [PubMed]
4. Grimme, S. Accurate description of van der Waals complexes by density functional theory including empirical corrections. *J. Comput. Chem.* **2004**, *25*, 1463–1473. [CrossRef] [PubMed]
5. Grimme, S. Semiempirical GGA-type density functional constructed with a long-range dispersion correction. *J. Comput. Chem.* **2006**, *27*, 1787–1799. [CrossRef] [PubMed]
6. Grimme, S.; Antony, J.; Ehrlich, S.; Krieg, H. A consistent and accurate ab initio parametrization of density functional dispersion correction (DFT-D) for the 94 elements H-Pu. *J. Chem. Phys.* **2010**, *132*, 154104. [CrossRef]
7. Goerigk, L.; Grimme, S. A thorough benchmark of density functional methods for general main group thermochemistry, kinetics, and noncovalent interactions. *Phys. Chem. Chem. Phys.* **2011**, *13*, 6670–6688. [CrossRef] [PubMed]
8. Grimme, S.; Ehrlich, S.; Goerigk, L. Effect of the damping function in dispersion corrected density functional theory. *J. Comput. Chem.* **2011**, *32*, 1456–1465. [CrossRef]
9. Caldeweyher, E.; Bannwarth, C.; Grimme, S. Extension of the D3 dispersion coefficient model. *J. Chem. Phys.* **2017**, *147*, 034112. [CrossRef] [PubMed]
10. Tkatchenko, A.; Scheffler, M. Accurate molecular van der waals interactions from ground-state electron density and free-atom reference data. *Phys. Rev. Lett.* **2009**, *102*, 073005 . [CrossRef]
11. Tkatchenko, A.; DiStasio, R.A.; Car, R.; Scheffler, M. Accurate and efficient method for many-body van der waals interactions. *Phys. Rev. Lett.* **2012**, *108*, 236402. [CrossRef]
12. Ruiz, V.G.; Liu, W.; Tkatchenko, A. Density-functional theory with screened van der waals interactions applied to atomic and molecular adsorbates on close-packed and non-close-packed surfaces. *Phys. Rev. B* **2016**, *93*, 035118. [CrossRef]
13. Dion, M.; Rydberg, H.; Schröder, E.; Langreth, D.C.; Lundqvist, B.I. Van der waals density functional for general geometries. *Phys. Rev. Lett.* **2004**, *92*, 246401. [CrossRef]
14. Lee, K.; Murray, D.; Kong, L.; Lundqvist, B.I.; Langreth, D.C. . Higher-accuracy van der waals density functional. *Phys. Rev. B* **2010**, *82*, 081101. [CrossRef]
15. Hyldgaard, P.; Berland, K.; Schröder, E. Interpretation of van der waals density functionals. *Phys. Rev. B* **2014**, *90*, 075148. [CrossRef]

16. Berland, K.; Cooper, V.R.; Lee, K.; Schroeder, E.; Thonhauser, T.; Hyldgaard, P.; Lundqvist, B.I. Van der Waals forces in density functional theory: A review of the vdW-DF method. *Rep. Prog. Phys.* **2015**,*78*, 066501. [CrossRef] [PubMed]

17. Klimeš, J.; Bowler, D.R.; Michaelides, A. Chemical accuracy for the van der waals density functional. *J. Phys. Condens. Matter* **2010**, *22*, 022201. [CrossRef]

18. Klimeš, J.; Bowler, D.R.; Michaelides, A. Van der waals density functionals applied to solids. *Phys. Rev. B* **2011**, *83*, 195131. [CrossRef]

19. Cooper, V.R. Van der waals density functional: An appropriate exchange functional. *Phys. Rev. B* **2010**, *81*, 161104. [CrossRef]

20. Berland, K.; Hyldgaard, P. Exchange functional that tests the robustness of the plasmon description of the van der waals density functional. *Phys. Rev. B* **2014**, *89*, 035412 [CrossRef]

21. Hamada, I. van der waals density functional made accurate. *Phys. Rev. B* **2014**, *89*, 121103. [CrossRef]

22. Peng, H.; Perdew, J.P. Rehabilitation of the perdew-burke-ernzerhof generalized gradient approximation for layered materials. *Phys. Rev. B* **2017**, *95*, 081105. [CrossRef]

23. Terentjev, A.V.; Constantin, L.A.; Pitarke, J.M. Dispersion-corrected pbesol exchange-correlation functional. *Phys. Rev. B* **2018**, *98*, 214108. [CrossRef]

24. Peng, H.; Yang, Z.; Perdew, J.P.; Sun, J. Versatile van der waals density functional based on a meta-generalized gradient approximation. *Phys. Rev. X* **2016**, *6*, 041005. [CrossRef]

25. Terentjev, A.V.; Cortona, P.; Constantin, L.A.; Pitarke, J.M.; Sala, F.D.; Fabiano, E. Solid-state testing of a van-der-waals-corrected exchange-correlation functional based on the semiclassical atom theory. *Computation* **2018**, *6*, 7. [CrossRef]

26. Csonka, G.I.; Perdew, J.P.; Ruzsinszky, A.; Philipsen, P.H.T.; Lebègue, S.; Paier, J.; Vydrov, O.A.; Ángyán, J.G. Assessing the performance of recent density functionals for bulk solids. *Phys. Rev. B* **2009**, *79*, 155107. [CrossRef]

27. Conroy, M.W.; Oleynik, I.I.; Zybin, S.V.; White, C.T. First-principles anisotropic constitutive relationships in beta-cyclotetramethylene tetranitramine (beta-HMX). *J. Appl. Phys.* **2008**, *104*, 053506. [CrossRef]

28. Landerville, A.C.; Conroy, M.W.; Budzevich, M.M.; Lin, Y.; White, C.T.; Oleynik, I.I. Equations of state for energetic materials from density functional theory with van der Waals, thermal, and zero-point energy corrections. *Appl. Phys. Lett.* **2010**, *97*, 251908. [CrossRef]

29. Cui, H.-L.; Ji, G.-F.; Chen, X.-R.; Zhang, Q.-M.; Wei, D.-W.; Zhao, F. Phase transitions and mechanical properties of octahydro-1,3,5,7-tetranitro-1,3,5,7-tetrazocine in different crystal phases by molecular dynamics simulation. *J. Chem. Eng. Data* **2010**, *5*, 3121–3129. [CrossRef]

30. Perdew, J.P.; Burke, K.; Ernzerhof, M. Generalized gradient approximation made simple. *Phys. Rev. Lett.* **1996**, *77*, 3865. [CrossRef]

31. Ceperley, D.M.; Alder, B.J. Ground state of the electron gas by a stochastic method. *Phys. Rev. Lett.* **1980**, *45*, 566–569. [CrossRef]

32. Gulans, A.; Puska, M.J.; Nieminen, R.M. Linear-scaling self-consistent implementation of the van der waals density functional. *Phys. Rev. B* **2009**, *79*, 201105. [CrossRef]

33. Román-Pérez, G.; Soler, J.M. Efficient implementation of a van der waals density functional: Application to double-wall carbon nanotubes. *Phys. Rev. Lett.* **2009**, *103*, 096102. [CrossRef] [PubMed]

34. Wu, J.; Gygi, F. A simplified implementation of van der waals density functionals for first-principles molecular dynamics applications. *J. Chem. Phys.* **2012**, *136*, 224107. [CrossRef] [PubMed]

35. Zhang, Y.; Yang, W. Comment on "generalized gradient approximation made simple". *Phys. Rev. Lett.* **1998**, *80*, 890. [CrossRef]

36. Puzder, A.; Dion, M.; Langreth, D.C. Binding energies in benzene dimers: Nonlocal density functional calculations. *J. Chem. Phys.* **2006**, *124*, 164105. [CrossRef]

37. Jurecka, P.; Sponer, J.; Cerny, J.; Hobza, P. Benchmark database of accurate (mp2 and ccsd(t) complete basis set limit) interaction energies of small model complexes, dna base pairs, and amino acid pairs. *Phys. Chem. Chem. Phys.* **2006**, *8*, 1985–1993. [CrossRef]

38. Vydrov, O.A.; Voorhis, T.V. Nonlocal van der waals density functional: The simpler the better. *J. Chem. Phys.* **2010**, *133*, 244103. [CrossRef]

39. Vydrov, O.A.; Voorhis, T.V. Nonlocal van der waals density functional made simple. *Phys. Rev. Lett.* **2009**, *103*, 063004. [CrossRef]

40. Vydrov, O.A.; Voorhis, T.V. Implementation and assessment of a simple nonlocal van der waals density functional. *J. Chem. Phys.* **2010**, *13*, 164113. [CrossRef]
41. Sabatini, R.; Gorni, T.; de Gironcoli, S. Nonlocal van der waals density functional made simple and efficient. *Phys. Rev. B* **2013**, *87*, 041108. [CrossRef]
42. Wellendorff, J.; Lundgaard, K.T.; Møgelhøj, A.; Petzold, V.; Landis, D.D.; Nørskov, J.K.; Bligaard, T.; Jacobsen, K.W. Density functionals for surface science: Exchange-correlation model development with bayesian error estimation. *Phys. Rev. B* **2012**, *85*, 23514. [CrossRef]
43. Kohn, W.; Sham, L.J. Self-consistent equations including exchange and correlation effects. *Phys. Rev.* **1965**, *140*, A1133–A1138. [CrossRef]
44. Perdew, J.P.; Ruzsinszky, A.; Csonka, G.I.; Vydrov, O.A.; Scuseria, G.E.; Constantin, L.A.; Zhou, X.; Burke, K. Restoring the density-gradient expansion for exchange in solids and surfaces. *Phys. Rev. Lett.* **2008**, *100*, 136406. [CrossRef]
45. Murray, E.D.; Lee, K.; Langreth, D.C. Investigation of exchange energy density functional accuracy for interacting molecules. *J. Chem. Theory Comput.* **2009**, *5*, 2754–2762. [CrossRef] [PubMed]
46. Perdew, J.P.; Wang, Y. Accurate and simple density functional for the electronic exchange energy: Generalized gradient approximation. *Phys. Rev. B* **1986**, *33*, 8800–8802. [CrossRef]
47. Becke, A.D. On the large-gradient behavior of the density functional exchange energy. *J. Chem. Phys.* **1986**, *85*, 7184–7187. [CrossRef]
48. Hamada, I.; Otani, M. Comparative van der waals density-functional study of graphene on metal surfaces. *Phys. Rev. B* **2010**, *82*, 153412. [CrossRef]
49. Choi, C.S.; Boutin, H.P. A study of crystal structure of beta-cyclotetramethylene tetranitramine by neutron diffraction. *Acta Crystallogr. Sect. B Struct. Sci.* **1970**, *B 26*, 1235. [CrossRef]
50. Herrmann, M.; Engel, W.; Eisenreich, N. Thermal-analysis of the phases of hmx using X-ray diffraction. *Z. Krist.* **1993**, *204*, 121–128.
51. Peng, Q.; Rahul; Wang, G.; Liu, G.R.; De, S. Structures, mechanical properties, equations of state, and electronic properties of β-hmx under hydrostatic pressures: A dft-d2 study. *Phys. Chem. Chem. Phys.* **2014**, *16*, 19972–19983, . [CrossRef]
52. Tao, J.; Zheng, F.; Gebhardt, J.; Perdew, J.P.; Rappe, A.M. Screened van der waals correction to density functional theory for solids. *Phys. Rev. Mater.* **2017**, *1*, 020802. [CrossRef]
53. Cady, H.H.; Larson, A.C.; Cromer, D.T. The crystal structure of α-HMX and a refinement of the structure of β-HMX. *Acta Crystallogr.* **1963**, *16*, 617–623. [CrossRef]
54. Choi, C.S.; Prince, E. The crystal structure of cyclotrimethylenetrinitramine. *Acta Crystallogr. Sect. B Struct. Sci.* **1972**, *B28*, 2857–2862. [CrossRef]
55. Millar, D.I.A.; Oswald, I.D.H.; Francis, D.J.; Marshall, W.G.; Pulham, C.R.; Cumming, A.S. The crystal structure of beta-rdx-an elusive form of an explosive revealed. *Chem. Commun.* **2009**, 562–564. [CrossRef] [PubMed]
56. Olinger, B.; Roof, B.; Cady, H.H. The linear and volume compression of beta-hmx and rdx to 9 gpa. In *Symposium (Int) on High Dynamic Pressures*; S INT COMP MIL DENS: Paris, France, 1978; pp. 3–8.
57. Golovina, N.I.; Titkov, A.N.; Raevskii, A.V.; Atovmyan, L.O. Kinetics and mechanism of phase transitions in the crystals of 2,4,6-trinitrotoluene and benzotrifuroxane. *J. Solid State Chem.* **1994**, *113*, 229–238. [CrossRef]
58. Cady, H.H.; Larson, A.C. Pentaerythritol tetranitrate II: Its crystal structure and transformation to PETN I: An algorithm for refinement of crystal structures with poor data. *Acta Crystallogr. Sect.* **1975**, *31*, 1864–1869. [CrossRef]
59. Cady, H.H. The crystal structure of *N*-methyl-N-2,4,6-tetranitroaniline (tetryl). *Acta Crystallogr.* **1967**, *23*, 601–609. [CrossRef]
60. Cady, H.H.; Larson, A.C. The crystal structure of 1,3,5-triamino-2,4,6-trinitrobenzene. *Acta Crystallogr.* **1965**, *18*, 485–496. [CrossRef]
61. Björkman, T.; Gulans, A.; Krasheninnikov, A.V.; Nieminen, R.M. Van der waals bonding in layered compounds from advanced density-functional first-principles calculations. *Phys. Rev. Lett.* **2012**, *108*, 235502. [CrossRef]
62. Geim, A.K.; Grigorieva, I.V. Van der Waals heterostructures. *Nature* **2013**, *499*, 419–425. [CrossRef]
63. Ambrosetti, A.; Ferri, N.; DiStasio, R.A.; Tkatchenko, A. Wavelike charge density fluctuations and van der waals interactions at the nanoscale. *Science* **2016**, *351*, 1171–1176. [CrossRef]

64. Peng, Q.; Rahul; Wang, G.; Liu, G.R.; Grimme, S.; De, S. Predicting Elastic Properties of β-HMX from First-principles calculations. *J. Phys. Chem. B* **2015**, *119*, 5896–5903. [CrossRef]

65. Kresse, G.; Hafner, J. Ab initio molecular dynamics for liquid metals. *Phys. Rev. B* **1993**, *47*, 558. [CrossRef]

66. Hammer, B.; Hansen, L.B.; Nørskov, J.K. Improved adsorption energetics within density-functional theory using revised perdew-burke-ernzerhof functionals. *Phys. Rev. B* **1999**, *59*, 7413–7421. [CrossRef]

67. Blöchl, P.E. Projector augmented-wave method. *Phys. Rev. B* **1994**, *50*, 17953–17979. [CrossRef]

68. Jones, R.O.; Gunnarsson, O. The density functional formalism, its applications and prospects. *Rev. Mod. Phys.* **1989**, *61*, 689–746. [CrossRef]

69. Ambrosetti, A.; Alfè, D.; DiStasio, R.A.; Tkatchenko, A. Hard numbers for large molecules: Toward exact energetics for supramolecular systems. *J. Phys. Chem. Lett.* **2014**, *5*, 849–855. [CrossRef]

70. Ambrosetti, A.; Reilly, A.M.; DiStasio, R.A.; Tkatchenko, A. Long-range correlation energy calculated from coupled atomic response functions. *J. Chem. Phys.* **2014**, *140*, 18A508. [CrossRef]

71. Reilly, A.M.; Tkatchenko, A. Seamless and accurate modeling of organic molecular materials. *J. Phys. Chem. Lett.* **2013**, *4*, 1028–1033. [CrossRef]

72. Ambrosetti, A.; Silvestrelli, P.L. Hidden by graphene—Towards effective screening of interface van der waals interactions via monolayer coating. *Carbon* 2018, *139*, 486–491. [CrossRef]

73. Ambrosetti, A.; Silvestrelli, P.L. Faraday-like screening by two-dimensional nanomaterials: A scale-dependent tunable effect. *J. Phys. Chem. Lett.* **2019**, *10*, 2044–2050. [CrossRef] [PubMed]

crystals

MDPI

Article

Insight into Physical and Thermodynamic Properties of X_3Ir (X = Ti, V, Cr, Nb and Mo) Compounds Influenced by Refractory Elements: A First-Principles Calculation

Dong Chen [1,2,3], Jiwei Geng [1], Yi Wu [2], Mingliang Wang [1,*] and Cunjuan Xia [2,*]

1   State Key Laboratory of Metal Matrix Composites, Shanghai Jiao Tong University, No. 800 Dongchuan Road, Shanghai 200240, China; chend@sjtu.edu.cn (D.C.); gengjiwei163@sjtu.edu.cn (J.G.)
2   School of Materials Science & Engineering, Shanghai Jiao Tong University, No. 800 Dongchuan Road, Shanghai 200240, China; eagle51@sjtu.edu.cn
3   Anhui Province Engineering Research Center of Aluminium Matrix Composites, Huaibei 235000, China
*   Correspondence: mingliang_wang@sjtu.edu.cn (M.W.); xiacunjuan@sjtu.edu.cn (C.X.); Tel.: +86-21-34202540 (M.W. & C.X.)

Received: 21 January 2019; Accepted: 14 February 2019; Published: 18 February 2019

Abstract: The effects of refractory metals on physical and thermodynamic properties of X_3Ir (X = Ti, V, Cr, Nb and Mo) compounds were investigated using local density approximation (LDA) and generalized gradient approximation (GGA) methods within the first-principles calculations based on density functional theory. The optimized lattice parameters were both in good compliance with the experimental parameters. The GGA method could achieve an improved structural optimization compared to the LDA method, and thus was utilized to predict the elastic, thermodynamic and electronic properties of X_3Ir (X = Ti, V, Cr, Nb and Mo) compounds. The calculated mechanical properties (i.e., elastic constants, elastic moduli and elastic anisotropic behaviors) were rationalized and discussed in these intermetallics. For instance, the derived bulk moduli exhibited the sequence of $Ti_3Ir < Nb_3Ir < V_3Ir < Cr_3Ir < Mo_3Ir$. This behavior was discussed in terms of the volume of unit cell and electron density. Furthermore, Debye temperatures were derived and were found to show good consistency with the experimental values, indicating the precision of our calculations. Finally, the electronic structures were analyzed to explain the ductile essences in the iridium compounds.

Keywords: Ir-based intermetallics; refractory metals; elastic properties; ab initio calculations

---

1. Introduction

Ir-based superalloys have received intensive interest in the last decades due to their high melting temperature as well as their improved strength, oxidation resistance and corrosion resistance at higher temperatures [1–5]. Consequently, these intermetallics can be deemed a suitable choice for high-temperature applications. For example, the cubic $L1_2$ intermetallic compounds Ir_3X (X = Ti, Zr, Hf, Nb and Ta) containing refractory elements could be proposed as "refractory superalloys" based on their higher melting points and superior mechanical properties at higher temperatures [3–5]. Terada et al. [6] conducted measurements on thermal properties (i.e., thermal conductivity and thermal expansion) from 300 to 1100 K, and found that the $L1_2$ Ir_3X (X = Ti, Zr, Hf, Nb and Ta) compounds were characterized by a larger thermal conductivity and a smaller thermal expansion. Chen et al. [7] exhibited the elastic constants and moduli of binary $L1_2$ Ir-based compounds at ground states by first-principles calculations, and reported the higher elastic moduli of these compounds together with their brittle characteristics in nature. Liu et al. [8] studied the elastic and thermodynamic properties of

Ir$_3$Nb and Ir$_3$V under varying pressure (0–50 GPa) and temperature (0–1200 K), and found that both compounds were stable without phase transformations.

Meanwhile, the typical refractory intermetallics should also include A15 cubic structure compounds with refractory metal elements. For example, Pan et al. [9] reported the mechanical and electronic properties of Nb$_3$Si using the first-principles method. By combining the first-principles method with quasi harmonic approximation, Papadimitriou et al. [10–12] critically investigated the mechanical and thermodynamic properties of Nb$_3$X (X = Al, Ge, Si and Sn). Chihi et al. [13] theoretically evaluated the elastic and thermodynamic properties of V$_3$X (X = Si, Ge and Sn) intermetallics utilizing the first-principles calculations. Jalborg et al. [14] determined the electronic structure of V$_3$Ir, V$_3$Pt and V$_3$Au using the self-consistent semi relativistic linear muffin-tin orbital (LMTO) band calculations. Paduani et al. [15] investigated the chemical bonding behavior and estimated the electron-phonon coupling constants of V$_3$X (X = Ni, Pd, Pt) by the full-potential linearized augmented-plane-wave (FP-LAPW) method.

Therefore, the Ir-based intermetallics with A15 crystal structure should also have the potential to be applied as structural materials. These compounds have been studied for their structural and electronic properties. For instance, Standanmann et al. [16] determined the lattice parameters of A15 X$_3$Ir (X = Ti, V, Cr, Nb and Mo) compounds. Meschel et al. [17] reported the experimental standard enthalpy of formation for V$_3$Ir. Paduani and Kuhnen [18] studied the band structure and Fermi surface of V$_3$Ir using the FP-LAPW method, and discussed Knight shift behavior in the compound. Paduani et al. [19] reported the electronic properties of Nb$_3$Ir via FP-LAPW calculations. Nevertheless, to our knowledge, the elastic and thermodynamic properties of X$_3$Ir (X = Ti, V, Cr, Nb and Mo) intermetallics have rarely been discussed.

This research is divided into the following parts. In the second section, the computational methods of X$_3$Ir (X = Ti, V, Cr, Nb and Mo) intermetallics are offered in detail. In the third section, the results and discussions are presented and discussed based on the effects of refractory metals on the physical and thermodynamic properties of X$_3$Ir compounds, including structural properties, elastic propertie, anisotropic behaviors, anisotropic sound velocities, Debye temperatures, and electronic structures. In the fourth section, the conclusions are drawn and presented in detail.

2. Materials and Methods

The first-principles calculations were performed using the CASTEP code, which is based on the pseudopotential plane-wave within density functional theory [20,21]. Using the ultrasoft pseudopotential [22] to model the ion-electron exchange-correlation, both the generalized gradient approximation (GGA) with the function proposed by Perdew, Burke and Ernzer (PBE) [23,24] and the local density approximation (LDA) with Ceperley–Alder form [25] were used. Additionally, the basis atom states were set as: Ti$3s^33p^63d^35s^2$, V$3s^23p^63d^35s^2$, Cr$3s^23p^63d^54s^1$, Nb$4s^24p^64d^45s^1$, Mo$4s^24p^64d^55s^1$ and Ir$5d^76s^2$. Through a series of tests, the cutoff energy of 400 eV was determined. In addition, a $10 \times 10 \times 10$ k-point mesh in the Brillouin zone was set for the special points sampling integration for the intermetallics. Both lattice constants and atom coordinates should be optimized via minimizing the total energy. Furthermore, the Brodyden–Fletcher–Goldfarb–Shanno (BFGS) minimization scheme was used for the geometric optimization [26,27]. Overall, the maximum stress has to be within 0.02 GPa, the maximum ionic force has to be within 0.01 eV/Å, the maximum ionic displacement has to be within 5.0×10^{-4} Å and the difference of the total energy has to be within 5.0×10^{-6} eV/atom for the geometrical optimization. Finally, the total energy and electronic structure were calculated, followed by cell optimization with a self-consistent field tolerance (5.0×10^{-7} eV/atom). Using the corrected tetrahedron Blöchl method, the total energies at equilibrium structures were derived [28].

3. Results

3.1. Structural Properties

The X_3Ir (X = Ti, V, Cr, Nb and Mo) intermetallics have a A15 cubic structure with the *cP8* (No. 223) space group. In a unit cell, six X atoms and two Ir atoms are dominated at the sites of 6c (0.25, 0, 0.5) and 2a (0, 0, 0), respectively. For the sake of performing structural property optimizations on IrX_3 compounds, the GGA method as well as the LDA method were utilized. The results are exhibited in Table 1, showing that the derived lattice constants using both methods are close to the experimental values [29–33].

Table 1. The optimized and experimental lattice parameters, the calculated deviations and densities for X_3Ir (X = Ti, V, Cr, Nb and Mo) compounds.

Compounds	a_0 (Å)	a_{exp} (Å)	Calculated Deviation (%)	Density (g/cm^3)
Ti$_3$Ir	5.010 [a]	5.012 [c]	−0.041 [a]	8.872 [a]
	4.901 [b]		−2.223 [b]	9.479 [b]
V$_3$Ir	4.7842 [a]	4.7876 [d]	−0.072 [a]	10.463 [a]
	4.6913 [b]		−2.012 [b]	11.099 [b]
Cr$_3$Ir	4.652 [a]	4.685 [e]	−0.712 [a]	11.489 [a]
	4.651 [b]		−0.732 [b]	11.496 [b]
Nb$_3$Ir	5.1585 [a]	5.135 [f]	0.457 [a]	11.394 [a]
	5.0777 [b]		−1.116 [b]	11.946 [b]
Mo$_3$Ir	4.9874 [a]	4.9703 [g]	0.344 [a]	12.986 [a]
	4.9199 [b]		−1.014 [b]	13.387 [b]

[a]: From the GGA method in this work: Theoretical values. [b]: From the LDA method in this work: Theoretical values. [c]: From Reference [29]: Experimental values. [d]: From Reference [30]: Experimental values. [e]: From Reference [31]: Experimental values. [f]: From Reference [32]: Experimental values. [g]: From Reference [33]: Experimental values.

In most compounds, the lattice constants generated by the GGA method offer much smaller calculated deviations than the LDA method (Table 1). For instance, the a_0(GGA) has the calculated deviation of −0.041%, and a_0(LDA) has the calculated deviation of −2.223% in comparison with a_{exp} in Ti$_3$Ir. Clearly, the GGA method exhibited better reliability for structural optimization, and thus giving a superior quality of calculation over the LDA method. As a result, the following calculation work was accomplished only by GGA method.

3.2. Elastic Constants

In the crystalline materials, the elastic constant represented the capability of resisting the exterior imposed stress. In such manners, a whole package of elastic constants was achieved to characterize mechanical properties of crystals. By imposing small strains to the equilibrium unit cell, elastic constants can be computed by determining the corresponding variations in the total energy. Theoretically, the elastic strain energy was formulated by Equation (1):

$$U = \frac{\Delta E}{V_0} = \frac{1}{2}\sum_i^6 \sum_j^6 C_{ij}e_i e_j \tag{1}$$

where V_0 represents the cell volume at equilibrium state; ΔE represents the energy difference; e_i and e_j represent the strains; C_{ij} (ij = 1, 2, 3, 4, 5 and 6) represent the elastic constants.

In cubic structures, C_{11}, C_{12} and C_{44} are nonzero elastic constants without mutual dependence. In Table 2, the calculated elastic constants (C_{ij}) for X_3Ir intermetallics are shown, accompanied by the available theoretical values [18,34–36] for comparison.

Table 2. The elastic constant (C_{ij}), Cauthy pressure (C_{12}-C_{44}), bulk modulus (B), shear modulus (G), Young's modulus (E), Poisson's ratio (v) and B/G ratio for X_3Ir (X = Ti, V, Cr, Nb and Mo) intermetallics.

Compounds	C_{ij}			C_{12}-C_{44} (GPa)	B (GPa)	G (GPa)	E (GPa)	v	B/G
	C_{11} (GPa)	C_{44} (GPa)	C_{12} (GPa)						
Ti$_3$Ir	183.8	52.8	166.0	114.2	171.9	26.5	75.7	0.427	6.483
	207.2 [a]	48.8 [a]	153.1 [a]	104.3 [a]	171.1 [a]	38.5 [a]	107.4 [a]	0.395 [a]	4.446 [a]
V$_3$Ir	471.5	109.4	136.8	27.4	248.4	129.8	331.7	0.277	1.913
					279.89 [b]				
Cr$_3$Ir	478.6	89.6	190.2	100.4	286.3	108.5	289.0	0.332	2.639
Nb$_3$Ir	433.7	84.5	123.7	39.2	227.0	108.0	279.7	0.295	2.102
					216.4 [c]				
Mo$_3$Ir	512.7	87.6	175.8	88.2	288.1	114.2	302.6	0.325	2.523
					297.5 [d]				

[a]: From Reference [34]: Theoretical values. [b]: From Reference [18]: Theoretical values. [c]: From Reference [35]: Theoretical values. [d]: From Reference [36]: Theoretical values.

In the elastic constant, a larger C_{44} corresponds to a stronger resistance to monoclinic shear in the (100) plane, and therefore symbolizes a larger shear modulus. For instance, V$_3$Ir has the largest C_{44} and shear modulus, exhibiting a superior capability to resist the shear stress. Furthermore, the compressive resistance along the *x* axis is reflected by C_{11}. For each compound, the derived C_{11} exhibited the biggest value among elastic constants, suggesting that is has the greatest incompressibility under *x* uniaxial stress. Among the compounds, Mo$_3$Ir was the least compressible along the *x* axis because it had the biggest C_{11} (512.7 GPa), and Ti$_3$Ir was the most compressible owing to its small C_{11} (183.8 GPa)

Utilizing Born's criteria [37,38], the essence of mechanical stability should be evaluated for cubic crystals:

$$C_{11} > 0; \; C_{44} > 0; \; C_{11} - C_{12} > 0; \; C_{11} + 2\,C_{12} > 0 \tag{2}$$

Using the values in Table 2, all X$_3$Ir compounds were found to have mechanical stability by satisfying the Born's criteria at the ground state.

The Cauchy pressure, illustrated as (C_{12}−C_{44}) [39], should be an effective indicator to evaluate the ductile/brittle nature of cubic crystals. In Pettifor's work [40], a more positive Cauchy pressure symbolized better ductility in the compound [41]. In Table 2, the Cauchy pressures for X$_3$Ir compounds were all positive in the order of V$_3$Ir < Nb$_3$Ir < Mo$_3$Ir < Cr$_3$Ir < Ti$_3$Ir, which means that X$_3$Ir compounds are naturally ductile. Such a result is in good compliance with the Cauchy pressure of Ti$_3$Ir provided by Rajagopalan [34]. Similarly, other A15 cubic crystals (i.e., V$_3$X (X = Si and Ge) [13], Nb$_3$X (X = Al, Ge, Si and Sn) [10] and Nb$_3$X (X = Al, Ga, In, Sn and Sb) [42]) have ductile characters owing to their positive Cauchy pressures.

3.3. Elastic Properties

Once the elastic constants were achieved, the elastic moduli (i.e., bulk modulus (B) and shear modulus (G)) could be computed by means of the Voigt–Reuss–Hill (VRH) method [43]. In cubic structures, the equations are exhibited as [44–46]:

$$B_V = B_R = \frac{1}{3}(C_{11} + 2C_{12}) \tag{3a}$$

$$G_V = \frac{1}{5}(C_{11} - C_{12} + 3C_{44}) \tag{3b}$$

$$G_R = \frac{5(C_{11} - C_{12})C_{44}}{4C_{44} + 3(C_{11} - C_{12})} \tag{3c}$$

$$B = \frac{B_V + B_G}{2} \tag{3d}$$

$$G = \frac{G_V + G_G}{2} \tag{3e}$$

When the elastic moduli are achieved, the Young's modulus (E) and Poisson's ratio (ν) should be calculated in the second step [47]:

$$E = \frac{9BG}{3B + G} \tag{3f}$$

$$\nu = \frac{3B - 2G}{2(3B + G)} \tag{3g}$$

Lastly, the computed elastic moduli for X_3Ir compounds using the VRH method are tabulated in Table 2 in combination with the available theoretical results for comparison [18,34–36]. Comparably, our calculated bulk moduli showed satisfactory agreement with the theoretical values for Ti_3Ir [34], Nb_3Ir [35] and Mo_3Ir [36], and a value slightly smaller smaller than the theoretical one for V_3Ir [18].

Analytically, the resisting capability against volume fluctuation under pressure is determined by the bulk modulus. For X_3Ir (X = Ti, V, Cr, Nb and Mo) intermetallics, the bulk moduli showed the sequence of $Ti_3Ir < Nb_3Ir < V_3Ir < Cr_3Ir < Mo_3Ir$. In References [48,49], a larger equilibrium cell volume was reported to correspond to a lower bulk modulus in the cubic crystal. Observably, such a conclusion is effective when the alloying elements are in the same cycle of the periodic table of elements (Figure 1a). For example, the bulk moduli are improved in the order of $Ti_3Ir < V_3Ir < Cr_3Ir$ depending on the reduced equilibrium cell volume. Similarly, the bulk modulus of Nb3Ir is smaller than that of Mo_3Ir with the larger equilibrium cell volume of Nb_3Ir (Figure 1a). Nevertheless, when the alloying elements are in the same group of the periodic table, the conclusion is valid for $Nb_3Ir < V_3Ir$, but ineffective for $Cr_3Ir < Mo_3Ir$, where Mo_3Ir actually has a larger equilibrium cell volume.

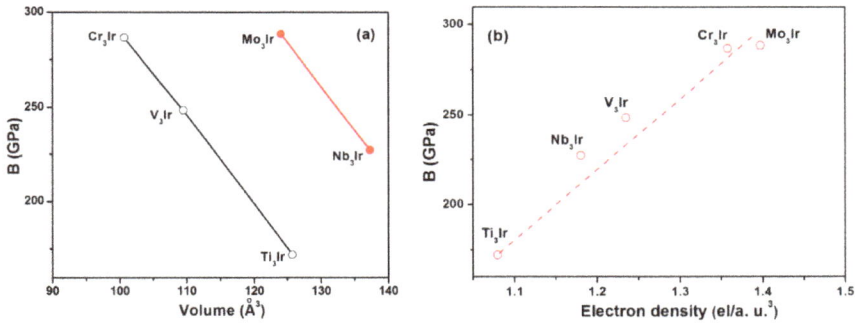

Figure 1. The relationship between bulk modulus and (**a**) volume of the unit cell or (**b**) electron density.

In order to further illustrate the relationship between the equilibrium cell volume and the bulk modulus of X_3Ir intermetallics, the linear dependence of the electron density on the bulk modulus is exhibited in Figure 1b. Clearly, dividing the bonding valence (ZB) by the volume per atom (VM) can deduce the electron density (n) in metallic compounds [50]. For X_3Ir compounds, the electron density (n) can be formulated as:

$$n(X_3Ir) = Z_B(X_3Ir)/V_M(X_3Ir) \tag{4a}$$

where VM(X_3Ir) represents the volume (cm^3/mol) of X_3Ir.

Rationalized by Vegard's law [51], ZB(X_3Ir) showed a bonding valence in (el/atom), and the Reference [52] tabulated the bonding valence of the pure element:

$$Z_B(X_3Ir) = (3Z_B(X) + Z_B(Ir))/4 \tag{4b}$$

Using this method, the linear dependence of the electron density on the bulk modulus was identified through the calculated values. Conclusively, it is more precise to rationalize the bulk modulus from the electron density, rather than the equilibrium cell volume.

Shear modulus (G) symbolizes the capability to resist shape fluctuation [44], and Young's modulus (E) is a measurement of resistance to tension and compression in the elastic regime [53]. Notably, there is a linear dependence of the Young's modulus on the shear modulus following the order of $Ti_3Ir <$ $Nb_3Ir < Cr_3Ir < Mo_3Ir < V_3Ir$ (Figure 2).

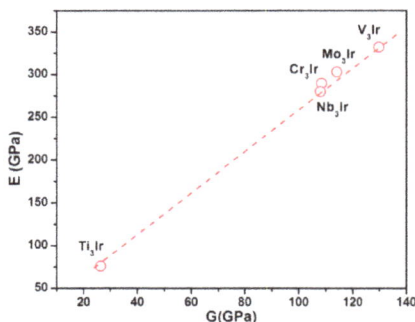

Figure 2. The relationship between shear modulus (G) and Young's modulus (E).

Overall, the bigger bulk modulus over shear modulus for each X_3Ir compound should reflect that the X_3Ir compound has an improved capability to resist volume fluctuation over shape fluctuation (Table 2). This conclusion complies well with the available data regarding the dependence of the bulk modulus on the shear modulus in other A15 intermetallics, i.e., Ti_3Ir (X = Ir, Pt and Au) [34], V_3X (X = Si and Ge) [13], Nb_3X (X = Al, Ga, In, Sn and Sb) [42] and Mo_3X (X = Si and Ge) [54].

The Poisson's ratio ($-1 \leq v \leq 0.5$) is used to quantify the stability of crystals against the shear deformation [55]. Materials with improved plasticity should possess a larger Poisson's ratio. The X_3Ir compounds have Poisson's ratios in the order of $V_3Ir < Nb_3Ir < Mo_3Ir < Cr_3Ir < Ti_3Ir$. This means that Ti_3Ir should be the most ductile, while V_3Ir is most brittle. Additionally, the Poisson's ratio provides information on the bonding forces in solids [56]. The lower and higher limits are 0.25 and 0.5 for the central force in a solid, respectively. For XIr_3 intermetallics, the interatomic forces of intermetallics should be central forces, since all the obtained values are located on this scale (Table 2).

The B/G ratio formulated by Pugh [57] is commonly adopted to quantitatively estimate the brittle or ductile essence of metallic compounds. The critical B/G ratio to distinguish the brittle from ductile material is 1.75. A smaller value is connected with the brittle nature, whereas a larger B/G ratio is related to ductility. Furthermore, the revised Cauchy pressure $((C_{12}-C_{44})/E)$ [58] was plotted against the B/G ratio to clarify the extent of ductility intuitively (Figure 3). As a result, the ductility was found to be enhanced in the order of $V_3Ir < Nb_3Ir < Mo_3Ir < Cr_3Ir < Ti_3Ir$. This conclusion agrees well with the analysis of the Poisson's ratio. Clearly, Ti_3Ir should be much more ductile than the other X_3Ir compounds (Figure 3).

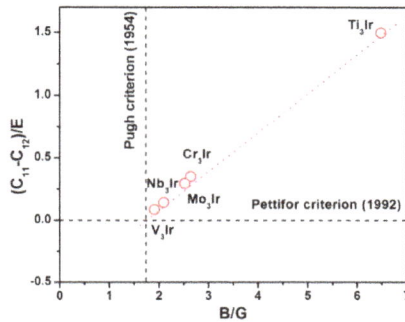

Figure 3. Revised Cauchy pressure $(C_{12}-C_{44})$/E as a factor of the B/G ratio for X_3Ir compounds.

3.4. Elastic Anisotropy

The universal anisotropic index (A^U) can be used to evaluate the elastic anisotropy, which is also referred to as the probability to introduce materials' micro-cracks [59]. The index can be calculated via Equation (5) [60]:

$$A^U = 5\frac{G_V}{G_R} + \frac{B_V}{B_R} - 6 \tag{5}$$

where B_V (B_R) and G_V (G_R) represent the symbols of the bulk modulus and the shear modulus at Voigt (Reuss) bounds, respectively.

In the calculated elastic anisotropies, B_V/B_R, should be equal to 1 for cubic crystals (Table 3).

Table 3. The computed bulk and shear moduli at Voigt (Reuss) bounds, and the universal anisotropic index (A^U) for X_3Ir compounds.

Compounds	B_V	B_R	G_V	G_R	B_V/B_R	G_V/G_R	A^U
Ti$_3$Ir	171.9	171.9	35.3	17.8	1	1.984	4.922
V$_3$Ir	248.4	248.4	132.6	127.0	1	1.044	0.220
Cr$_3$Ir	286.3	286.3	111.4	105.6	1	1.055	0.277
Nb$_3$Ir	227.0	227.0	112.7	103.3	1	1.091	0.455
Mo$_3$Ir	288.1	288.1	119.9	108.4	1	1.106	0.532

Indeed, G_V/G_R has a decisive effect on the universal anisotropic index (A^U). Figure 4 shows that the universal anisotropic index increases linearly with the increment of the G_V/G_R value. A compound with a smaller A^U represents a weaker extent of anisotropy. Therefore, the universal anisotropy was found to be reduced in the sequence of V$_3$Ir < Cr$_3$Ir < Nb$_3$Ir < Mo$_3$Ir < Ti$_3$Ir. Generally, Ti$_3$Ir has the largest universal anisotropy, and V$_3$Ir has the smallest. Because the experimental value is lacking for comparison in these compounds, this calculation has to be evaluated in later research.

In addition, to further describe the anisotropy of X_3Ir compounds, the directional dependence of the reciprocal of the Young's modulus was constructed for a three-dimensional (3D) surface according to Equation (6) [48]:

$$\frac{1}{E} = S_{11} - 2(S_{11} - S_{12} - \frac{S_{44}}{2})(l_1^2 l_2^2 + l_2^2 l_3^2 + l_1^2 l_3^2) \tag{6}$$

where S_{ij} represents the usual elastic compliance constant obtained from the inverse of the matrix of the elastic constant; l_1, l_2 and l_3 represent the direction cosines in the sphere coordination.

If a crystal has ideal isotropic performance, the 3D directional dependence of the Young's modulus would show a spherical shape. In fact, the extent of deviation from the spherical shape symbolizes the anisotropic extent. In Figure 5, X_3Ir compounds showed the distinctive 3D figures of Young's moduli with various deviations from a sphere. This confirmed that X_3Ir compounds have anisotropic behaviors. Obviously, Ti_3Ir shows the largest deviation from the sphere shape along the <111> direction. On the contrary, other X_3Ir compounds exhibited different forms of deviation, and the most visible deviations were observed along the zone axes. Finally, the extent of the elastic anisotropy for X_3Ir obeyed the sequence of $V_3Ir < Cr_3Ir < Nb_3Ir < Mo_3Ir < Ti_3Ir$. This conclusion complies well with the result obtained from the universal anisotropic index.

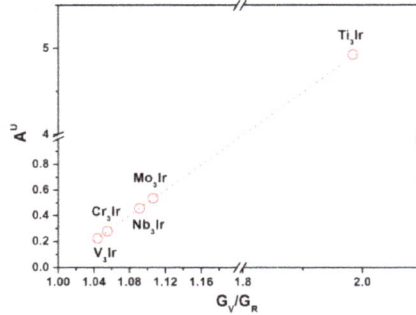

Figure 4. The correlation between G_V/G_R and the universal anisotropic index (A^U).

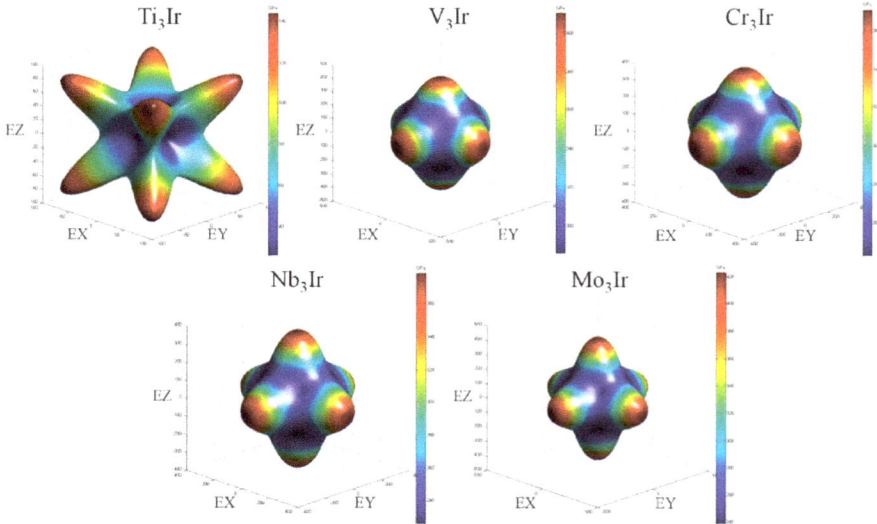

Figure 5. The 3D surface construction of the Young's modulus in X_3Ir compounds. (The magnitudes of Young's moduli at different directions are presented by the contours along each graph with the unit of GPa).

3.5. Anisotropic Sound Velocity and Debye Temperature

In the crystalline material, the sound velocities should depend on the crystalline symmetry in combination with the propagating direction. In the cubic structure, [111], [110] and [001] directions exhibited the pure transverse and longitudinal modes, accordingly. Regarding other directions, both quasi-transverse and quasi-longitudinal waves can work as the main sound propagating modes. Therefore, the sound velocities formulated in the principal directions are listed as follows [61]:

$$
\begin{aligned}
&[100]v_l = \sqrt{C_{11}/\rho}; [010]v_{t1} = [001]v_{t2} = \sqrt{C_{44}/\rho}\\
&[110]v_l = \sqrt{(C_{11}+C_{12}+C_{44})/(2\rho)}\\
&[1\bar{1}0]v_{t1} = \sqrt{(C_{11}-C_{12})/\rho}; [001]v_{t2} = \sqrt{C_{44}/\rho}\\
&[111]v_l = \sqrt{(C_{11}+2C_{12}+4C_{44})/(3\rho)}\\
&[11\bar{2}]v_{t1} = [11\bar{2}]v_{t2} = \sqrt{(C_{11}-C_{12}+C_{44})/(3\rho)}
\end{aligned}
\tag{7}
$$

where v_l (v_t) represents the longitudinal (transverse) sound velocity; ρ represents the density (see Table 1).

Overall, the longitudinal sound velocity along the [100] direction was only decided by C_{11}. The transverse modes along [010] and [001] directions were dependent on C_{44}. The longitudinal sound velocities along both the [110] and [111] directions were influenced by C_{11}, C_{12} and C_{44}.

Along the [100], [110] and [111] directions, the longitudinal sound velocities and the transverse sound velocities are exhibited in Table 4 for each X_3Ir compound. For each compound, the longitudinal sound velocity followed the rising sequence of [100] < [110] < [111]. The sound velocities showed anisotropic properties, further confirming the elastic anisotropic behaviors of the compounds.

These theoretically computed physical properties (i.e., elastic moduli and Poisson's ratio) and structural properties (i.e., density) should be adopted to calculate the Debye temperature (Θ) using the following formula [54,62,63]:

$$
\Theta = \frac{h}{k}\left[\frac{3n}{4\pi}\left(\frac{N_A\rho}{M}\right)\right]^{\frac{1}{3}} v_D
\tag{8a}
$$

where ρ represents the density (see Table 1); h represents the Planck's constant ($h = 6.626 \times 10^{-34}$ J/s); k represents the Boltzmann's constant ($k = 1.381 \times 10^{-23}$ J/K); n represents the number of atoms per formula unit; N_A represents the Avogadro's number ($N_A = 6.023 \times 10^{-23}$/mol); M represents the molecular weight (M(Ti$_3$Ir) = 335.8 g/mol, M(V$_3$Ir) = 345 g/mol, M(Cr$_3$Ir) = 347.8 g/mol, M(Nb$_3$Ir) = 470.9 g/mol, M(Mo$_3$Ir) = 480 g/mol); v_D represents the average sound velocity in polycrystalline materials. The latter is formulated as:

$$
v_D = \left[\frac{1}{3}\left(\frac{1}{V_L^3} + \frac{2}{V_T^3}\right)\right]^{-\frac{1}{3}}
\tag{8b}
$$

where v_T and v_L represent the transverse and longitudinal sound velocities, respectively, as formulated by the equations below:

$$
v_T = \sqrt{\frac{G}{\rho}}
\tag{8c}
$$

$$
v_L = \sqrt{\frac{B+\frac{4}{3}G}{\rho}}
\tag{8d}
$$

Table 4. The anisotropic sound velocities (m/s), average sound velocities (m/s) and Debye temperatures (K) for X_3Ir intermetallics.

Crystalline Orientation		Ti_3Ir	V_3Ir	Cr_3Ir	Nb_3Ir	Mo_3Ir
[111]	$[111]v_l$	5226.6	6138.7	5942.8	5460.3	5583.6
	$[11\bar{2}]v_{t1,2}$	1629.2	3762.0	3311.6	3397.4	3301.0
[110]	$[110]v_l$	4763.3	5856.6	5744.8	5307.3	5466.2
	$[1\bar{1}0]v_{t1}$	1416.9	5656.7	5010.3	5216.1	5093.6
	$[001]v_{t2}$	2440.3	3234.2	2792.4	2723.7	2597.2
[100]	$[100]v_l$	4551.4	6713.4	6454.1	6169.4	6283.3
	$[010]v_{t1}$	2440.3	3234.2	2792.4	2723.7	2597.2
	$[001]v_{t2}$	2440.3	3234.2	2792.4	2723.7	2597.2
	v_L	4833.4	6346.8	6124.7	5706.4	5822.9
	v_T	1728.8	3522.6	3073.2	3079.0	2965.2
	v_D	1962.7	3920.4	3444.0	3434.1	3320.0
	Θ	233.3	487.9	441.0	396.4	397.8
		238 [a], 262.6 [b]	460 ± 10 [c], 445 [d]	449 [e]	409 ± 8 [c], 377 [d]	452 [f], 325 [g], 497.06 [h]

[a]: From Reference [64]: Experimental values. [b]: From Reference [34]: Theoretical values. [c]: From Reference [65]: Experimental values. [d]: From Reference [66]: Experimental values. [e]: From Reference [67]: Experimental values. [f]: From Reference [16]: Experimental values. [g]: From Reference [68]: Experimental values. [h]: From Reference [36]: Theoretical values.

For each X_3Ir compound, the derived Debye temperature (Θ) is tabulated in Table 4. Also, the published experimental [16,64–68] and theoretical [34,36] values are included for comparison.

Generally, V_3Ir had the largest Debye temperature, and Ti_3Ir had the smallest. The calculated Debye temperatures were in the order of $Ti_3Ir < Nb_3Ir < Mo_3Ir < Cr_3Ir < V_3Ir$. Clearly, our results revealed the reduced tendency of Debye temperatures with the M atom in the same group, i.e., Cr (Lighter element) and Mo (Heavier element) are from Group-VIB, and V (Lighter element) and Nb (Heavier element) are from Group-VB.

Comparably, the obtained Debye temperatures for Ti_3Ir, V_3Ir, Cr_3Ir and Nb_3Ir were all in excellent agreement with the available experimental results [16,64–68]. For instance, the calculated Θ was 233 K for Ti_3Ir. This agrees well with the experimental values reported by Junod et al. [64] with the calculated deviation of 2.01%. Notably, the Debye temperature reported by Rajagopalan et al. [34] had the calculated deviation of 10.3%, indicating the poor quality of the prediction in this work. However, for Mo_3Ir, the published experimental [16,68] and theoretical [36] Debye temperatures were quite scattered, although our calculated values were closer to the experimental values from Staudenmann's report [16]. Nevertheless, more works are required on this compound.

Because both structural parameters and elastic moduli are incorporated to calculate the Debye temperature, the superior quality of our calculation on these structural and elastic parameters using the GGA method was evidenced by the smaller differences between the estimated and experimental values of Debye temperatures.

3.6. Electronic Structures

Figure 6a–e exhibit the density of states (DOS) spectra representing the calculated electronic structures for X_3Ir compounds. The DOS spectra for these A15 cubic phases were similar to each other. In a typical DOS spectrum, there are normally three regions, including the lower electron band, the upper electron band, and the conduction unoccupied states around the Fermi level (E_F). For example (Figure 6a), the lower the electron band was mainly contributed by $4s$ electrons of Ti ranging from -55 to -57.5 eV. The upper electron band was occupied by $3p$ electrons of Ti ranging from -32 to -35 eV. Around the Fermi level, the conduction unoccupied states were created through the hybridization of mainly Ti$3d$ electrons with Ti$3p$, Ir$5d$ and Ir$4p$ electrons. X_3Ir compounds were plotted around the

Fermi level at zero in all the total DOS (TDOS) and partial DOS (PDOS) spectra. Clearly, no any energy gap can be found near the Fermi level. Therefore, their nature of metallicity was confirmed.

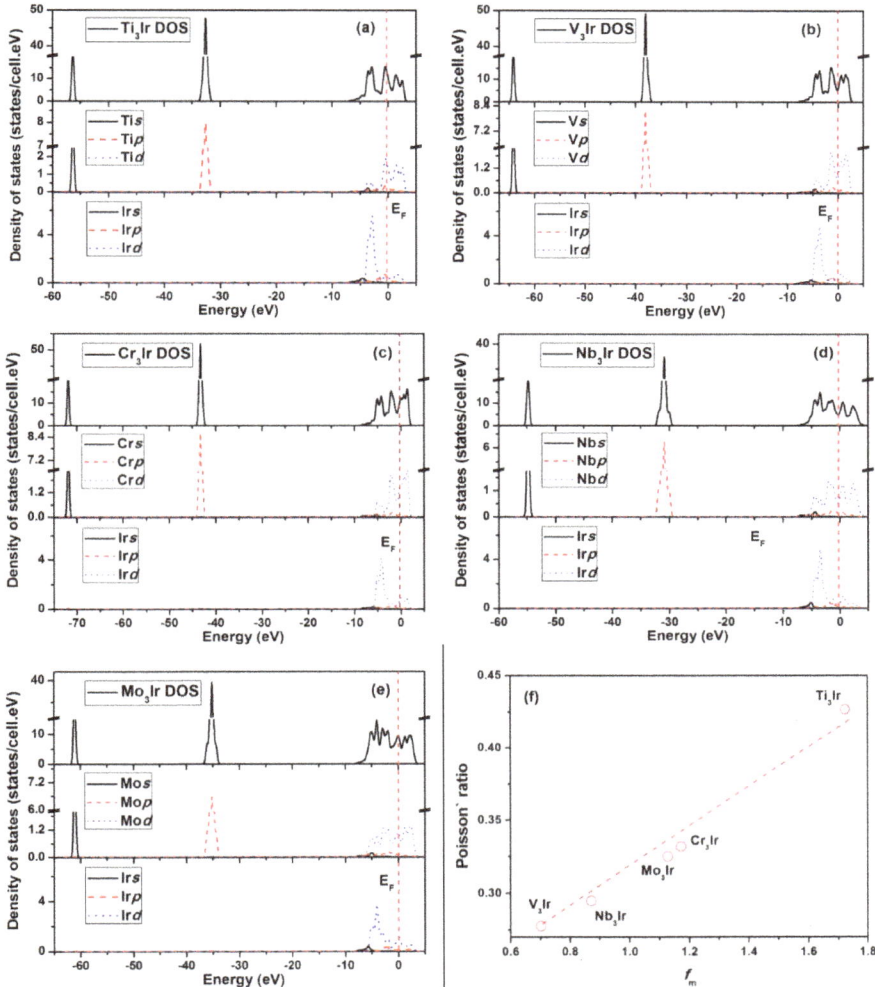

Figure 6. Total density of states (TDOS) and partial density of states (PDOS) spectra for (a) Ti₃Ir, (b) Cr₃Ir, (c) Mo₃Ir, (d) Nb₃Ir and (e) V₃Ir; (f) the correlation between metallicity and the Poisson's ratio in X₃Ir compounds.

Furthermore, the electron density values can provide quantitative evidence of the metallic nature in the bonding characteristics. Even at the Fermi surface, the electron density values were much larger than zero. According to the electronic Fermi liquid theory [69], the metallicity of the compound has to be estimated using Equation (9) [70]:

$$f_m = \frac{n_m}{n_e} = \frac{k_B T D_f}{n_e} = \frac{0.026 D_f}{n_e} \tag{9}$$

where

k_B represents the Boltzmann constant ($k = 1.381 \times 10^{-23}$ J/K);

T represents the absolute temperature;

D_f represents the DOS value at the Fermi level;

n_m and n_e represent the thermally excited electrons and valence electron density of the cell, respectively; n_e is calculated by $n_e = N/V_{cell}$ (N represents the total number of valence electrons; V_{cell} represents the cell volume).

Using the calculated metallicity values (f_m), the correlation between metallicity and Poisson's ratios can be constructed for X_3Ir compounds (Figure 6f). It was found that the Poisson's ratios were diminished with the reduction in metallicity in compounds with the order of $V_3Ir < Nb_3Ir < Mo_3Ir < Cr_3Ir < Ti_3Ir$. This indicated that a compound with higher metallicity in its bonds should possess better ductility.

4. Conclusions

The effects of refractory metals on physical and thermodynamic properties of X_3Ir (X = Ti, V, Cr, Nb and Mo) intermetallics were investigated utilizing first-principles calculations. The conclusions are listed as follows:

(1) Using the GGA method to structurally optimized the unit cell, smaller calculation deviations for lattice constants were achieved as compared to those achieved using the LDA method.

(2) The calculated bulk moduli exhibited the increasing sequence of $Ti_3Ir < Nb_3Ir < V_3Ir < Cr_3Ir < Mo_3Ir$. Furthermore, the bulk moduli showed a linear relationship with electron densities. The Young's modulus showed a linear dependence on shear modulus following the order of $Ti_3Ir < Nb_3Ir < Cr_3Ir < Mo_3Ir < V_3Ir$.

(3) Based on the discussions on the Cauchy pressure, Poisson's ratio and B/G ratio, the ductile essence was found to be enhanced in the order of $V_3Ir < Nb_3Ir < Mo_3Ir < Cr_3Ir < Ti_3Ir$.

(4) For X_3Ir compounds, the extent of the elastic anisotropy for X_3Ir obeyed the increasing sequence of $V_3Ir < Cr_3Ir < Nb_3Ir < Mo_3Ir < Ti_3Ir$ via the analyses of the universal anisotropic indexes and 3D surface constructions.

(5) The Debye temperatures obtained for Ti_3Ir, V_3Ir, Cr_3Ir and Nb_3Ir were all in good agreement with the results from experiments. Such good compliance proved the superior quality of our calculations of the structural and elastic properties, since the computation of Debye temperature is concerned with both structural and elastic parameters.

(6) The calculated electronic structures for X_3Ir compounds showed similar features in the DOS spectra. Furthermore, the metallicity of the compounds was calculated, and was correlated with the Poisson's ratios. This indicated that a compound with higher metallicity in its bonds should possess better ductility.

Author Contributions: In this work, D.C. has contributed on the conceptualization, investigation, data curation, formal analysis, and writing—original draft preparation. J.G. has contributed on the investigation, data curation, and writing—review and editing. Y.W. has contributed on the formal analysis, and writing—review and editing and funding acquisition. M.W. has contributed on the conceptualization, investigation, data curation, formal analysis, and writing—review and editing. C.X. has contributed on the conceptualization, investigation, formal analysis, writing—review and editing and funding acquisition.

Funding: This work was sponsored by the National Key Research and Development Program of China (Grant No. 2018YFB1106302) and the project (Grant No. 2017WAMC002) sponsored by Anhui Province Engineering Research Center of Aluminum Matrix Composites (China).

Conflicts of Interest: The authors have declared no conflict of interest.

References and Notes

1. Yamaguchi, M.; Inui, H.; Ito, K. High-temperature structural intermetallics. *Acta Mater.* **2000**, *48*, 307–322. [CrossRef]
2. Yu, X.; Yamabe-Mitarai, Y.; Ro, Y.; Harada, H. New developed quaternary refractory superalloys. *Intemetallics* **2000**, *8*, 619–622. [CrossRef]
3. Yamabe-Mitarai, Y.; Gu, Y.; Huang, C.; Völkl, R.; Harada, H. Platinum-group-metal-based intermetallics as high-temperature structural materials. *JOM* **2004**, *56*, 34–39. [CrossRef]
4. Yamabe-Mitari, Y.; Ro, Y.; Maruko, T.; Harada, H. Microstructure dependence of strength of Ir-base refractory superalloys. *Intermetallics* **1999**, *7*, 49–58. [CrossRef]
5. Yamabe-Mitarai, Y.; Ro, Y.; Nakazawa, S. Temperature dependence of the flow stress of Ir-based $L1_2$ intermetallics. *Intermetallics* **2001**, *9*, 423–429. [CrossRef]
6. Terada, Y.; Ohkubo, K.; Miura, S.; Sanchez, J.M.; Mohri, T. Thermal conductivity and thermal expansion of Ir_3X (X = Ti, Zr, Hf, V, Nb, Ta) compounds for high-temperature applications. *Mater. Chem. Phys.* **2003**, *80*, 385–390. [CrossRef]
7. Chen, K.; Zhao, L.R.; Tse, J.S. Ab initio study of elastic properties of Ir and Ir_3X compounds. *J. Appl. Phys.* **2003**, *93*, 2414–2417. [CrossRef]
8. Liu, N.; Wang, X.; Wan, Y. Firstprinciple calculations of elastic and thermodynamic properties of Ir3Nb and Ir3V with L12 structure under high pressure. *Intermetallics* **2015**, *66*, 103–110. [CrossRef]
9. Pan, Y.; Lin, Y.; Xue, Q.; Ren, C.; Wang, H. Relationship between Si concentration and mechanical properties of Nb-Si compounds: A first-principles study. *Mater. Des.* **2016**, *89*, 676–683. [CrossRef]
10. Papadimitriou, I.; Utton, C.; Scott, A.; Tsakiropoulos, P. Ab Initio Study of Binary and Ternary $Nb_3(X,Y)$ A15 Intermetallic Phases (X,Y = Al, Ge, Si, Sn). *Metall. Mater. Trans. A* **2015**, *46*, 566–576. [CrossRef]
11. Papadimitriou, I.; Utton, C.; Tsakiropoulos, P. Ab initio investigation of the intermetallics in the Nb-Sn binary system. *Acta Mater.* **2015**, *86*, 23–33. [CrossRef]
12. Papadimitriou, I.; Utton, C.; Tsakiropoulos, P. Ab initio investigation of the Nb-Al system. *Comput. Mater. Sci.* **2015**, *107*, 116–121. [CrossRef]
13. Chihi, T.; Fatmi, M. Theoretical prediction of the structural, elastic, electronic and thermodynamic properties of V_3M (M = Si, Ge and Sn) compounds. *Superlatt. Microstruc.* **2012**, *52*, 697–703. [CrossRef]
14. Jarlborg, T.; Junod, A.; Peter, M. Electronic structure, superconductivity, and spin fluctuations in the A15 compounds A_3B: A =V, Nb; B=Ir, Pt, Au. *Phys. Rev. B* **1983**, *27*, 1558–1567. [CrossRef]
15. Paduani, C.; Kuhnen, C.A. Band structure calculations in isoelectronic V3B compounds: B=Ni, Pd and Pt. *Solid State Commun.* **2010**, *150*, 1303–1307. [CrossRef]
16. Staudenmann, J.L.; DeFacio, B.; Testardi, L.R.; Werner, S.A.; Flükiger, R.; Muller, J. Debye classes in A15 compounds. *Phys. Rev. B* **1981**, *24*, 6446. [CrossRef]
17. Meschel, S.V.; Kleppa, O.J. The standard enthalpies of formation of some intermetallic compounds of transition metals by high temperature direct synthesis calorimetry. *J. Alloy. Compd.* **2006**, *415*, 143–149. [CrossRef]
18. Paduani, C.; Kuhnen, C.A. Electronic structure of A15-type compounds: V_3Co, V_3Rh, V_3Ir and V_3Os. *Eur. Phys. J. B* **2009**, *69*, 331–336. [CrossRef]
19. Paduani, C. Electronic properties of the A-15 Nb-based intermetallics Nb3(Os,Ir,Pt,Au). *Solid State Commun.* **2007**, *144*, 352–356. [CrossRef]
20. Segall, M.D.; Lindan, P.J.D.; Probert, M.J.; Pickard, C.J.; Hasnip, P.J.; Clark, S.J.; Payne, M.C. First-principles simulation: Ideas, illustrations and the CASTEP code. *J. Phys. Condens. Matter* **2002**, *14*, 2717–2744. [CrossRef]
21. Clark, S.J.; Segall, M.D.; Pickard, C.J.; Hasnip, P.J.; Probert, M.J.; Refson, K.; Payne, M.C. First principles methods using CASTEP. *Zeitschrift fuer Kristallographie* **2005**, *220*, 567–570. [CrossRef]
22. Vanderbilt, D. Soft self-consistent pseudopotentials in a generalized eigenvalue formalism. *Phys. Rev. B* **1990**, *41*, 7892–7895. [CrossRef]
23. Perdew, J.P.; Wang, Y. Accurate and simple analytic representation of the electron-gas correlation energy. *Phys. Rev. B* **1992**, *45*, 13244–13249. [CrossRef]
24. Perdew, J.P.; Burke, K.; Ernzerhof, M. Generalized Gradient Approximation Made Simple. *Phys. Rev. Lett.* **1996**, *77*, 3865–3868. [CrossRef]

25. Perdew, J.P.; Zunger, A. Self-interaction correction to density-functional approximations for many-electron systems. *Phys. Rev. B* **1981**, *23*, 5048–5079. [CrossRef]

26. Shanno, D.F. Conditioning of quasi-Newton methods for function minimization. *Math. Comput.* **1970**, *24*, 647–656. [CrossRef]

27. Fischer, T.H.; Almlof, J. General Methods for Geometry and Wave Function Optimization. *J. Chem. Phys.* **1992**, *96*, 9768–9774. [CrossRef]

28. Blöchl, P.E.; Jepsen, O.; Andersen, O.K. Improved tetrahedron method for Brillouin-zone integrations. *Phys. Rev. B* **1994**, *49*, 16223–16233. [CrossRef]

29. Ti_3Ir, ICSD No. 641115.

30. V_3Ir, ICSD No. 104591.

31. Cr_3Ir, ICSD No. 102780.

32. Nb_3Ir, ICSD No. 640833.

33. Mo_3Ir, ICSD No.600761.

34. Rajagopalan, M.; Gandhi, R.R. First principles study of structural, electronic, mechanical and thermal properties of A15 intermetallic compounds Ti_3X (X = Au, Pt, Ir). *Phys. B* **2012**, *407*, 4731–4734. [CrossRef]

35. Paduani, C. Structural and electronic properties of the A-15 compounds Nb_3Rh and Nb_3Ir. *Phys. B* **2007**, *393*, 105–109. [CrossRef]

36. Subhashree, G.; Sankar, S.; Krithiga, R. Superconducting properties of Mo_3Os,Mo_3Pt,Mo_3Ir from first principle calculations. *Mod. Phys. Lett. B* **2014**, *28*, 1450233. [CrossRef]

37. Afaq, A.; Rizwan, M.; Bakar, A. Computational investigations of XMgGa (X = Li, Na) half Heusler compounds for thermo-elastic and vibrational properties. *Phys. B* **2019**, *554*, 102–106. [CrossRef]

38. Heciri, D.; Belkhir, H.; Belghit, R.; Bouhafs, B.; Khenata, R.; Ahmed, R.; Bouhemadou, A.; Ouahrani, T.; Wang, X.; Omrani, S.B. Insight into the structural, elastic and electronic properties of tetragonal inter-alkali metal chalcogenides CsNaX (X = S, Se, and Te) from first-principles calculations. *Mater. Chem. Phys.* **2019**, *221*, 125–137. [CrossRef]

39. He, D.G.; Lin, Y.C.; Jiang, X.Y.; Yin, L.X.; Wang, L.H.; Wu, Q. Dissolution mechanisms and kinetics of δ phase in an aged Ni-based superalloy in hot deformation process. *Mater. Des.* **2018**, *156*, 262–271. [CrossRef]

40. Pettifor, D.G. Theoretical predictions of structure and related properties of intermetallics. *Mater. Sci. Technol.* **1992**, *8*, 345–349. [CrossRef]

41. Fatima, B.; Chouhan, S.S.; Acharya, N.; Sanyal, S.P. Theoretical prediction of the electronic structure, bonding behavior and elastic moduli of scandium intermetallics. *Intermetallics* **2014**, *53*, 129–139. [CrossRef]

42. Sundareswari, M.; Ramasubramanian, S.; Rajagopalan, M. Elastic and thermodynamical properties of A15 Nb_3X (X = Al,Ga,In,Sn and Sb) compounds-First principles DFT study. *Solid State Commun.* **2010**, *150*, 2057–2060. [CrossRef]

43. Han, Y.; Wu, Y.; Li, T.; Khenata, R.; Yang, T.; Wang, X. Electronic, Magnetic, Half-Metallic, and Mechanical Properties of a New Equiatomic Quaternary Heusler Compound YRhTiGe: A First-Principles Study. *Materials* **2018**, *11*, 797. [CrossRef]

44. Chen, D.; Chen, Z.; Wu, Y.; Wang, M.; Ma, N.; Wang, H. First-principles investigation of mechanical, electronic and optical properties of Al_3Sc intermetallic compound under pressure. *Comput. Mater. Sci.* **2014**, *91*, 165–172. [CrossRef]

45. Salma, M.U.; Rahman, M.A. Study of structural, elastic, electronic, mechanical, optical and thermodynamic properties of $NdPb_3$ intermetallic compound: DFT based calculations. *Comput. Condens. Matter* **2018**, *15*, 42–47. [CrossRef]

46. Luan, X.; Qin, H.; OrcID, F.L.; Dai, Z.; Yi, Y.; Li, Q. The Mechanical Properties and Elastic Anisotropies of Cubic Ni_3Al from First Principles Calculations. *Crystals* **2018**, *8*, 307. [CrossRef]

47. Sultana, F.; Uddin, M.M.; Ali, M.A.; Hossain, M.M.; Naqib, S.H.; Islam, A.K.M.A. First principles study of M_2InC (M = Zr, Hf and Ta) MAX phases: The effect of M atomic species. *Results Phys.* **2018**, *11*, 869–876. [CrossRef]

48. Chen, S.; Sun, Y.; Duan, Y.H.; Huang, B.; Peng, M.J. Phase stability, structural and elastic properties of C15-type Laves transition-metal compounds MCo_2 from first-principles calculations. *J. Alloy. Compd.* **2015**, *630*, 202–208. [CrossRef]

49. Li, C.X.; Duan, Y.H.; Hu, W.-C. Electronic structure, elastic anisotropy, thermal conductivity and optical properties of calcium apatite $Ca_5(PO_4)_3X$ (X = F, Cl or Br). *J. Alloy. Compd.* **2015**, *619*, 66–77. [CrossRef]

50. Huang, S.; Zhang, C.H.; Li, R.Z.; Shen, J.; Chen, N.X. Site preference and alloying effect on elastic properties of ternary B2 RuAl-based alloys. *Intermetallics* **2014**, *51*, 24–29. [CrossRef]
51. Jacob, K.T.; Raj, S.; Rannesh, L. Vegard's law: A fundamental relation or an approximation? *Inter. J. Mater. Res.* **2007**, *98*, 776–779. [CrossRef]
52. Li, C.; Wu, P. Correlation of Bulk Modulus and the Constituent Element Properties of Binary Intermetallic Compounds. *Chem. Mater.* **2001**, *13*, 4642–4648. [CrossRef]
53. Chen, D.; Chen, Z.; Wu, Y.; Wang, M.; Ma, N.; Wang, H. First-principles study of mechanical and electronic properties of TiB compound under pressure. *Intermetallics* **2014**, *52*, 64–71. [CrossRef]
54. Zhong, S.Y.; Chen, Z.; Wang, M.; Chen, D. Structural, elastic and thermodynamic properties of Mo_3Si and Mo_3Ge. *Eur. Phys. J. B* **2016**, *89*, 6. [CrossRef]
55. Lebga, N.; Daoud, S.; Sun, X.W.; Bioud, N.; Latreche, A. Mechanical and Thermophysical Properties of Cubic Rock-Salt AlN Under High Pressure. *J. Electr. Mater.* **2018**, *47*, 3430–3439. [CrossRef]
56. Fu, H.; Li, D.; Peng, F.; Gao, T.; Cheng, X. Ab initio calculations of elastic constants and thermodynamic properties of NiAl under high pressures. *Comput. Mater. Sci.* **2008**, *44*, 774–778. [CrossRef]
57. Pugh, S.F. XCII. Relations between the elastic moduli and the plastic properties of polycrystalline pure metals. *Philos. Mag.* **1954**, *45*, 823–843. [CrossRef]
58. Niu, H.Y.; Chen, X.Q.; Liu, P.T.; Xing, W.W.; Cheng, X.Y.; Li, D.Z.; Li, Y.Y. Extra-electron induced covalent strengthening and generalization of intrinsic ductile-to-brittle criterion. *Sci. Rep.* **2012**, *2*, 718. [CrossRef]
59. Han, C.; Chai, C.; Fan, Q.; Yang, J.; Yang, Y. Structural, Electronic, and Thermodynamic Properties of Tetragonal t-SixGe3-xN4. *Materials* **2018**, *11*, 397. [CrossRef]
60. Belghit, R.; Belkhir, H.; Kadri, M.T.; Heciri, D.; Bououdin, M.; Ahuja, R. Structural, elastic, electronic and optical properties of novel antiferroelectric KNaX (X = S, Se, and Te) compounds: First principles study. *Phys. B* **2018**, *545*, 18–29. [CrossRef]
61. Duan, Y.H.; Sun, Y.; Peng, M.J.; Zhou, S.G. Anisotropic elastic properties of the Ca-Pb compounds. *J. Alloy. Compd.* **2014**, *595*, 14–21. [CrossRef]
62. Vajeeston, P.; Ravindran, P.; Fjellvag, H. Prediction of structural, lattice dynamical, and mechanical properties of CaB_2. *RSC Adv.* **2012**, *2*, 11687–11694. [CrossRef]
63. Haque, E.; Hossain, M.A. First-principles study of elastic, electronic, thermodynamic, and thermoelectric transport properties of TaCoSn. *Results Phys.* **2018**, *10*, 458–465. [CrossRef]
64. Junod, A.; Flukiger, R.; Muller, J. Supraconductivite et chaleur specifique dans les alliages A15 a base de titane. *J. Phys. Chem. Solids* **1976**, *37*, 27–31. [CrossRef]
65. Spitzli, P. Chaleur spécifique d'alliages de structure A 15. *Physik der kondensierten Materie* **1971**, *13*, 22–58.
66. Junod, A.; Bischel, D.; Muller, J. Eliashberg inversion of superconducting state thermodynamics. *Helv. Phys. Acta* **1979**, *52*, 580. [CrossRef]
67. Flükiger, R.; Heiniger, F.; Junod, A.; Muller, J.; Spitzli, P.; Staudenmann, J.L. Chaleur specifique et supraconductivite dans des alliages de structure A 15 a base de chrome. *J. Phys. Chem. Solids* **1971**, *32*, 459–463. [CrossRef]
68. Morin, F.J.; Maita, J.P. Specific Heats of Transition Metal Superconductors. *Phys. Rev.* **1963**, *129*, 1115. [CrossRef]
69. Misawa, S. The 3-Dimensional Fermi Liquid Description for the Iron-Based Superconductors. *J. Low Temp. Phys.* **2018**, *190*, 45–66. [CrossRef]
70. Li, Y.; Gao, Y.; Xiao, B.; Min, T.; Fan, Z.; Ma, S.; Xu, L. Theoretical study on the stability, elasticity, hardness and electronic structures of W-C binary compounds. *J. Alloy. Compd.* **2010**, *502*, 28–37. [CrossRef]

![crystals logo]

crystals

MDPI

Article

First-Principles Assessment of the Structure and Stability of 15 Intrinsic Point Defects in Zinc-Blende Indium Arsenide

Qing Peng [1,*], Nanjun Chen [1], Danhong Huang [2], Eric R. Heller [3], David A. Cardimona [2] and Fei Gao [1,*]

[1] Nuclear Engineering and Radiological Sciences, University of Michigan, Ann Arbor, MI 48109, USA; njchen@umich.edu

[2] Space Vehicles Directorate, Air Force Research Laboratory, Kirtland AFB, Albuquerque, NM 87117, USA; danhong.huang@us.af.mil (D.H.); david.cardimona@us.af.mil (D.A.C.)

[3] Materials and Manufacturing Directorate, Air Force Research Laboratory, Wright Patterson AFB, OH 45433, USA; e-eric.heller@wpafb.af.mil

[*] Correspondence: qing@qpeng.org (Q.P.); gaofeium@umich.edu (F.G.)

Received: 4 December 2018; Accepted: 15 January 2019; Published: 17 January 2019

Abstract: Point defects are inevitable, at least due to thermodynamics, and essential for engineering semiconductors. Herein, we investigate the formation and electronic structures of fifteen different kinds of intrinsic point defects of zinc blende indium arsenide (zb-InAs) using first-principles calculations. For As-rich environment, substitutional point defects are the primary intrinsic point defects in zb-InAs until the n-type doping region with Fermi level above 0.32 eV is reached, where the dominant intrinsic point defects are changed to In vacancies. For In-rich environment, In tetrahedral interstitial has the lowest formation energy till n-type doped region with Fermi level 0.24 eV where substitutional point defects In_{As} take over. The dumbbell interstitials prefer $< 110 >$ configurations. For tetrahedral interstitials, In atoms prefer 4-As tetrahedral site for both As-rich and In-rich environments until the Fermi level goes above 0.26 eV in n-type doped region, where In atoms acquire the same formation energy at both tetrahedral sites and the same charge state. This implies a fast diffusion along the $t - T - t$ path among the tetrahedral sites for In atoms. The In vacancies V_{In} decrease quickly and monotonically with increasing Fermi level and has a $q = -3e$ charge state at the same time. The most popular vacancy-type defect is V_{In} in an As-rich environment, but switches to V_{As} in an In-rich environment at light p-doped region when Fermi level below 0.2 eV. This study sheds light on the relative stabilities of these intrinsic point defects, their concentrations and possible diffusions, which is expected useful in defect-engineering zb-InAs based semiconductors, as well as the material design for radiation-tolerant electronics.

Keywords: point defects; formation energy; indium arsenide; first-principles; charged defects

---

1. Introduction

The III-V zinc-blende semiconductors are among the most important semiconductors, and have recently received much attention since they have potential to be employed as base materials for light-emitting diodes, infrared photodetectors, and spintronic devices, e.g., quantum-dot and quantum-well applications [1–3]. The materials have been the subject of interest in a large variety of experimental and theoretical investigations [4–6]. The III-V semiconductors are strong candidates to be incorporated into high-performance opto-electronics due to their direct band gap and high electron mobility [7,8]. In the family of III-V materials, InAs stands out because of its very high electron mobility which can be as much as three times higher than those in InGaAs and GaAs [7,9]. Meanwhile, it also acquires a small direct band gap of 0.35 eV at room temperature and a low carrier effective masses as well [10]. Together, these properties make InAs a promising candidate for incorporation into next-generation nano-electronics [7]. Additionally, InAs has already been made successfully into nanowires [11–13] and demonstrated to integrate well into novel field-effect transistor (FET) device geometries [14–17].

Due to its important applications in electronics, extensive efforts have been put in studying the electronic properties and performance of nanowire-based devices [18–20] and quantum dots [21]. It is also desirable to understand the instabilities in these materials and devices under severe conditions including radiation damage and their survivability under single event upset [22,23]. One of the fundamental questions is the formation energy of point defects, which is essential to understand the creation of defects from an energetics aspect. Moreover, thermodynamic arguments suggest that the intrinsic or native defects will be inevitably present within a crystal at finite temperatures.

Under ambient conditions, InAs crystallizes into a cubic zinc-blende (zb) geometry with space group $F\bar{4}3m$ (T_d^2) [24]. The atomic structure of pristine zb-InAs in a conventional eight-atom unit cell is depicted in Figure 1a, and we will limit our study to this type of crystalline structure. In this computational study, we primarily focus on the formation energy of various point defects, providing insights in understanding defect energetics within the bulk InAs crystal. For enhancing accuracy, our calculations are performed at the Ab initio level using density-functional theory (DFT). Generally, DFT describes reasonably well the structural properties, such as lattice constants and bulk moduli [25]. For a more accurate description of defect structures, we have carried out the investigation with enlarged simulation cells (216 lattice sites systems) to ensure that the accuracy is within 0.02 eV/cell. A $3 \times 3 \times 3$ supercell with 108 In and 108 As atoms and referenced for defect calculations is illustrated in Figure 1b. Our study aims to provide an extensive and accurate study of the intrinsic point defect formation which is missed in the literature, e.g., a very recent computational study [26], but is highly desirable.

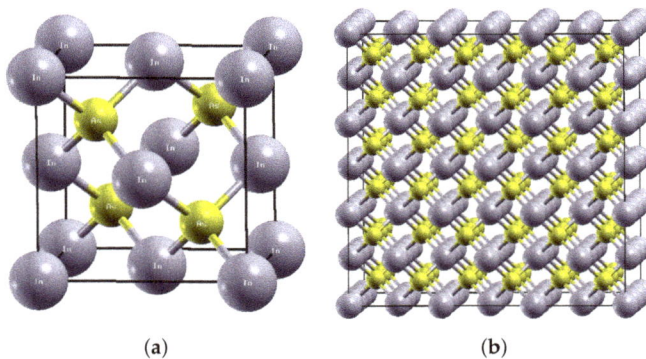

(a)　　　　　　　　　　(b)

Figure 1. Atomic structures of pristine zinc blende InAs in (**a**) conventional unit cell with 4 In and 4 As atoms and (**b**) $3 \times 3 \times 3$ super unit cell with 108 In and 108 As atoms, referenced for defect calculations. Here, the small yellow ball denotes As atoms, while the large silver balls are for In atoms.

The remainder of this paper is organized as follows. Section 2 presents the computational method, including the formula for defect formation energy and the finite-size corrections, as well as the details of DFT calculations. The results and analysis are presented in Section 3, discussing the Fermi-level dependence of the formation energies of fifteen intrinsic defects in five groups under different chemical environments and in various charge states. The conclusions are provided in Section 4.

2. Methods

2.1. Defect Formation Energy

To reduce the artificial self-image interactions imposed by periodic-boundary conditions, defects are generally modeled in an enlarged super cell. The selection of the super-cell size is a compromise between the accuracy and the computing demands. In general, the formation energy for a defect with charges in a semiconductor or an insulator has contributions from both ions and electrons. In a super-cell formalism, for a defect or impurity X in a charge state q, its formation energy $E^f(X^q)$ is computed by

$$E^f(X^q, E_F) = E^{tot}(X^q) - E^{tot}_{bulk}(q = 0)$$
$$- \sum_i \Delta n_i \mu_i + q(E_{VBM} + E_F) + E^{corr} , \qquad (1)$$

where $E^{tot}(X^q)$ is the total energy of the super-cell containing the defect X in the charge state q, $E^{tot}_{bulk}(q = 0)$ denotes the total energy of the pristine bulk supercell which is neutral and free of any defects, Δn_i represents the number of atoms of species i added to ($\Delta n_i > 0$) or removed from ($\Delta n_i < 0$) the supercell as a result of the defect formation, and $\mu_i = \mu_i^{bulk} + \Delta \mu_i$ corresponds to the chemical potential of element species i. When an atom is added to the system, the associated electrons are also added to the system and contribute to the formation energy. Such a contribution is described by the chemical potential of electrons, known as the Fermi level E_F at zero temperature. Here, E_F of a semiconductor is treated as an independent variable that can take any value within the bandgap. It is worth noting that E_F is measured with respect to E_{VBM}, the energy of valence band maximum (VBM) of the bulk material.

Since the exact value of the chemical potential μ_i in Eqaution (1) cannot be determined, it is treated as a parameter for the formation energy calculations. As such, the defect formation energies are given in the limiting conditions of As-rich and In-rich growth regimes. In the As-rich regime, the chemical potential of As atoms is assumed to be the value in bulk As, whereas in the In-rich (As-poor) regime, it corresponds to the chemical-potential difference between InAs and bulk In (and vice versa for the chemical potential of In). For an in-depth discussion of formation-energy calculations, the reader is referred to the following papers [27–30].

The correction term E^{corr} in Equation (1) is used to remove the errors introduced by finite size (L) effects and periodic-boundary conditions, such as spurious overlaps of neighboring defect wave functions and, in case of charged defects, Coulomb interactions between image charges. There is still extensive debate on the performance and applicability of different schemes of corrections [31–33], e.g., Makov and Payne (MP) scheme [34], alignment-only scheme [35], Freysoldt, Neugebauer and Van de Walle (FNV) scheme [36], Lany and Zunger (LZ) scheme [37]. The mutual relation between various schemes and defining the conditions for their applications are discussed by Komsa et al. [38]. The classical MP scheme is adopted in this paper and gives the correction terms as

$$E^{corr}(q, L) = E^{corr}_{mono} + E^{corr}_{quad} = -\frac{\alpha q^2}{\epsilon L} + \frac{A_3}{L^3} . \qquad (2)$$

Here, the first term is the monopole Madelung term [32], while the second term is the third-order quadrupole electrostatic correction. In addition, α is the Madelung constant of the crystal, q is the charge of a defect state, embedded in a uniform compensating background charge, with the unit of e (the positive electron charge). ϵ is the static dielectric constant. The third-order parameter A_3 is taken as a fitting parameter.

Several early studies indicated that the quadrupole correction does not always improve results, leaving its utility somewhat in question [39,40]. Therefore, in this paper we only consider the leading term of the monopole Madelung correction. The Madelung constant α is 1.638 for zinc blende cubic lattice of point charges, and 2.8373 for simple cubic lattice of point charges. Within the approximation of a single charge monopole, we adopt $\alpha = 2.8373$ for the cubic system. $\epsilon = 15.15$ is chosen for the static dielectric constant of InAs.

2.2. Details of Density Functional Theory Calculations

The conventional unit cell contains eight atoms (four In and four As atoms). We used a $3 \times 3 \times 3$ supercell containing 216 regular lattice sites, which consists of 108 In and 108 As atoms before including defects. The total energies of the system and forces on each atom are characterized via first-principles calculations within the framework of DFT. All DFT calculations are carried out with the Vienna Ab-initio Simulation Package (VASP) [41,42] which is based on the Kohn-Sham density-functional theory (KS-DFT) [43,44] with generalized gradient approximation for the exchange-correlation functionals [45] as parameterized by Perdew, Burke and Ernzerhof (PBE) and revised for solids (PBEsol) [46]. The electrons explicitly included in the calculations are the $4d^{10}5s^25p^1$ electrons (13 electrons) and the $4s^24p^3$ electrons (5 electrons) for each of In and As atom, respectively. The core electrons are replaced by the projector augmented wave (PAW) and pseudo-potential approach [47,48]. A plane-wave cutoff of 400 eV is used in the geometry relaxation to reduce Pulay stress. For all other calculations, we employ a plane-wave cutoff of 240 eV with accurate and dense k-mesh, where the irreducible Brillouin Zone is sampled with a Gamma-centered $3 \times 3 \times 3$ k-mesh. Moreover, the calculations were performed at zero temperature. The criterion for stopping the relaxation of the electronic degrees of freedom is set by requiring the total energy change to be smaller than 10^{-5} eV. The optimized atomic geometry is achieved through minimizing Hellmann-Feynman forces acting on each atom until the maximum forces on ions become smaller than 0.01 eV/Å.

After the geometry optimization of a perfect crystal, we introduce defects by either removal of an appropriate atom to create a vacancy, or addition of an extra atom to create an interstitial in a pre-specified position (tetrahedral or dumbbell). The resulting structures were allowed to relax energetically, permitting atoms to move in all three dimensions. Here, all geometry optimizations were performed using the classical conjugate gradient algorithm.

It is well known that GGA level exchange correlation functions severely underestimate the band gaps of semiconductors, although the the total energy of the system could be obtained with fairly and satisfactorily accurate [49]. More accurate bandgap prediction require higher level exchange correlation functionals including GW and HSE methods [25]. However, because the defective systems are in general contain hundreds of atoms, for example, about 216 atoms in this study, it is unfeasible to carry out higher levels (GW, or HSE for example) calculations. Therefore, here we used the experimental (0.417 eV) of the band gap throughout this study, as previous efforts [31,49,50].

3. Results and Analysis

3.1. Atomic Structures of Intrinsic Defects

Under the ambient condition, InAs, one of the most important III-V semiconductors, has a cubic 3C zinc-blende crystalline configuration, We first optimized the geometry of the pristine *zb*-InAs with lattice parameters measured between 6.0584 Å and 6.060 Å [51]. Our result with the lattice parameter of 6.058 Å from GGA-PBEsol agrees well with previous GGA-PBE results of 6.059 Å [20], PBE-PW91 results of 6.061 Å [52], and experiment of 6.0588 Å reported by Thompson, Rowe, and Bubenstein in 1969 [53].

We then generated fifteen different defect configurations for intrinsic point defects in *zb*-InAs , as depicted in Figure 2. Each defect configuration sits around the center of a $3 \times 3 \times 3$ supercell with 216 lattice sites, which ensures that the interactions between images become negligible. All these defect configurations are fully relaxed so that the maximum amplitude of the forces on every atom is less than 0.02 eV/Å. The final relaxed atomic structures of these fifteen configurations in neutral states (charge $q = 0$) are displayed in Figure 2. For denotation of these point defects, we take the form of the element or vacancy (V) with a subscript of the site. For example, As_{110} means a configuration with an As atom on the <110> dumbbell position, and In_{As} means a configuration with an In atom substituting for an As atom on the site.

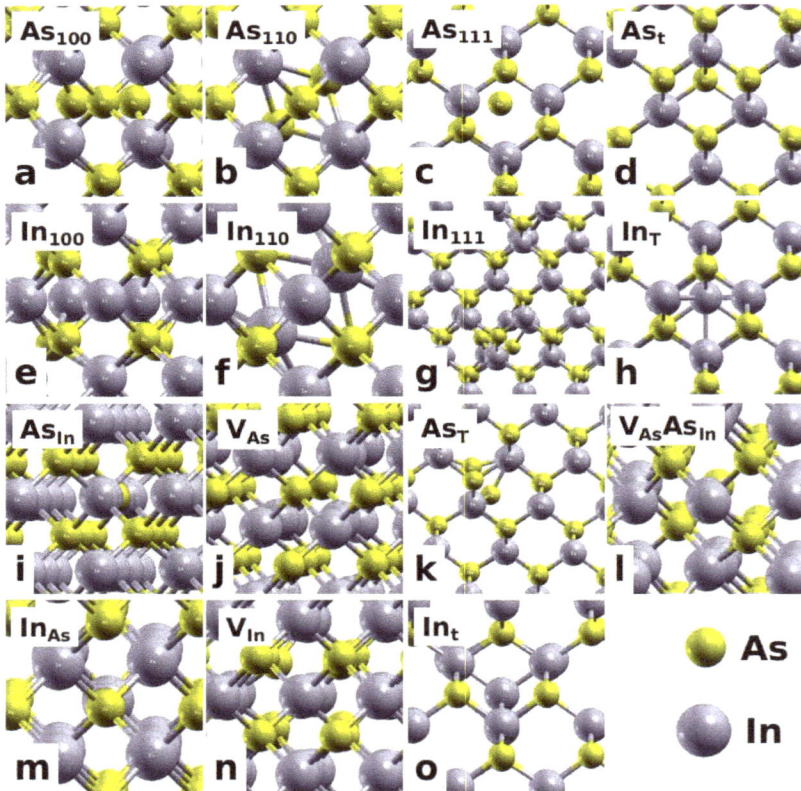

Figure 2. The atomic structures of 15 intrinsic point-defect configurations of *zb*-InAs after full geometry optimization according to the minimum energy in neural charge state: (**a**) As_{100}, (**b**) As_{110}, (**c**) As_{111}, (**d**) As_t, (**e**) In_{100}, (**f**) In_{110}, (**g**) In_{111}, (**h**) In_T, (**i**) As_{In}, (**j**) V_{As}, (**k**) As_T, (**l**) $V_{As}As_{In}$, (**m**) In_{As}, (**n**) V_{In}, (**o**) In_t. Here, the small yellow spheres denote As atoms, while the large silver spheres denotes In atoms.

The defects in Figure 2 are divided into five groups. Group-1 is the dumbbell interstitial-type point defects. The dumbbell configuration is characterized by two atoms of the same species sharing one lattice site. There are three typical orientations: <100> , <110> , and <111> . Therefore, there exist totally six dumbbell configurations: (a) As_{100}, (b) As_{110}, (c) As_{111}, and (e) In_{100}, (f) In_{110}, (g) In_{111} as seen in Figure 2. Group-2 is the tetrahedral interstitial-type point defects. There are two kinds of tetrahedral sites in a *zb*-InAs lattice: one is formed by four In atoms denoted as t site, and the other is formed by four As atoms denoted as T site. Both In and As atoms could take either atom site, resulting in four tetrahedral interstitial-type point defects: (d) As_t, (h) In_T, (k) As_T, and (o) In_t as presented in Figure 2. Group-3 is the substitutional point defects, where a pristine As site is replaced by an In atom, or *vise versa*. The two intrinsic substitutional point defects in *zb*-InAs are (i) As_{In} and (m) In_{As} as shown in Figure 2. Group-4 is the vacancy-type point defects. There are two types of intrinsic vacancy-type point defects in *zb*-InAs : (j) V_{As} and (n) V_{In}, as displayed in Figure 2. The last Group-5 is a point-defect complex such as $V_{As}As_{In}$, which is formed by an As vacancy V_{As} combined with a substitutional As atom on the nearest-neighbor In site (As$_{In}$), as demonstrated by Figure 2o. This defect is interesting because an In vacancy V$_{In}$ could attract a nearby As atom to fill it up, resulting in the V$_{As}$As$_{In}$ complex. It is worth pointing out that the relaxed atomic structures of the dumbbell interstitial-type defect of In$_{111}$ and tetrahedral interstitial-type defect of As$_T$ are also point-defect complexes, indicating they are highly unstable. Next, we investigate the formation energies of all these different types of defects.

3.2. Defect Formation Energies

The formation energies of defects are computed according to Equation (1) for different charge states from $q = -4$ to $q = 4$. In order to estimate the uncertainties caused by the finite size of supercells as well as the spurious image interactions, we therefore evaluate the corrections, as detailed in Section 2. As a result, we find the deviations up to 0.01 eV for defects with a charge of $q = \pm4$. We regard these values as unavoidable uncertainties which nevertheless do not significantly alter the main conclusions.

It is clear from the formula in Equation (1) that the formation energies of intrinsic defects will depend on the choice of chemical potentials, i.e., the choice of reservoir with which equilibrium is achieved. The chemical potentials are constrained by equilibrium conditions, which varies from case to case, location to location, time to time. Here we only consider two extreme conditions: (1) As-rich and (2) In-rich. Our results of the chemical potentials are computed as $\mu_{As}^{bulk} = -3.014$ eV, $\mu_{In}^{bulk} = -2.389$ eV, and $\mu_{InAs}^{bulk} = -6.574$ eV.

Another factor we need to consider is the Fermi level E_F, which is required for counting the formation energies from the electrons' contribution. However, the exact value of E_F is very sensitive to local environments including doping concentrations. Consequently, we have expressed the defect formation energy as a function of E_F which varies in the whole range of the electronic band gap. The experimental value of the bandgap is 0.417 [51]. We adopted this value as the range for E_F in our study. In fact, the formation energy depends linearly on the Fermi level, as manifested in Equation (1). The band-gap predicted from PBE solid is 0.142 eV, which is severely underestimated from the experimental value of 0.417 eV, because the artificial self-interaction of electrons are not excluded [54]. Here we have applied the "extended gap scheme" to map the calculated transfer levels onto the experimental band gap [39]. The regime near the edge of the valence bands or VBM is the *p*-type doped where E_F is small. The regime near the edge of the conduction bands or CBM is the *n*-type doped and E_F is large and comparable to the band gap. The two two different regimes are dislayed in the figures of the formation energies to mark the doping stage through out this study.

The formation energies of fifteen defect configurations in the dilute limit are computed using the supercell method as aforementioned. It is worth mentioning that although the point defects are generated in a non-equilibrium process, the relative formation energies of different configurations can shed light on preferential locations and determine the accessible ground-state charge states, as well as charge-state transition levels [54]. The formation energies as functions of E_F in the As-rich and In-rich environments for all these intrinsic defects in *zb*-InAs with all possible charge states from -4 to $+4$ are displayed in the upper and lower panels of Figure 3, respectively. Our results show the distinctive trend and significant variations of the formation energies under various charge states as functions of E_F for each point defect in both As-rich and In-rich environments. All calculated formation energies lie between 1.0 eV and 9.0 eV. The general trend is that the formation energy decreases for negative charge state ($q < 0$) but increases for positive charge state ($q > 0$) with increasing E_F. The slope of the formation energy as a function of E_F is positively correlated with the charge state. For $q = +4e$, the amount of the increment in formation energy is more than 1.5 eV as E_F changes from VBM (0 eV) to CBM (0.417 eV).

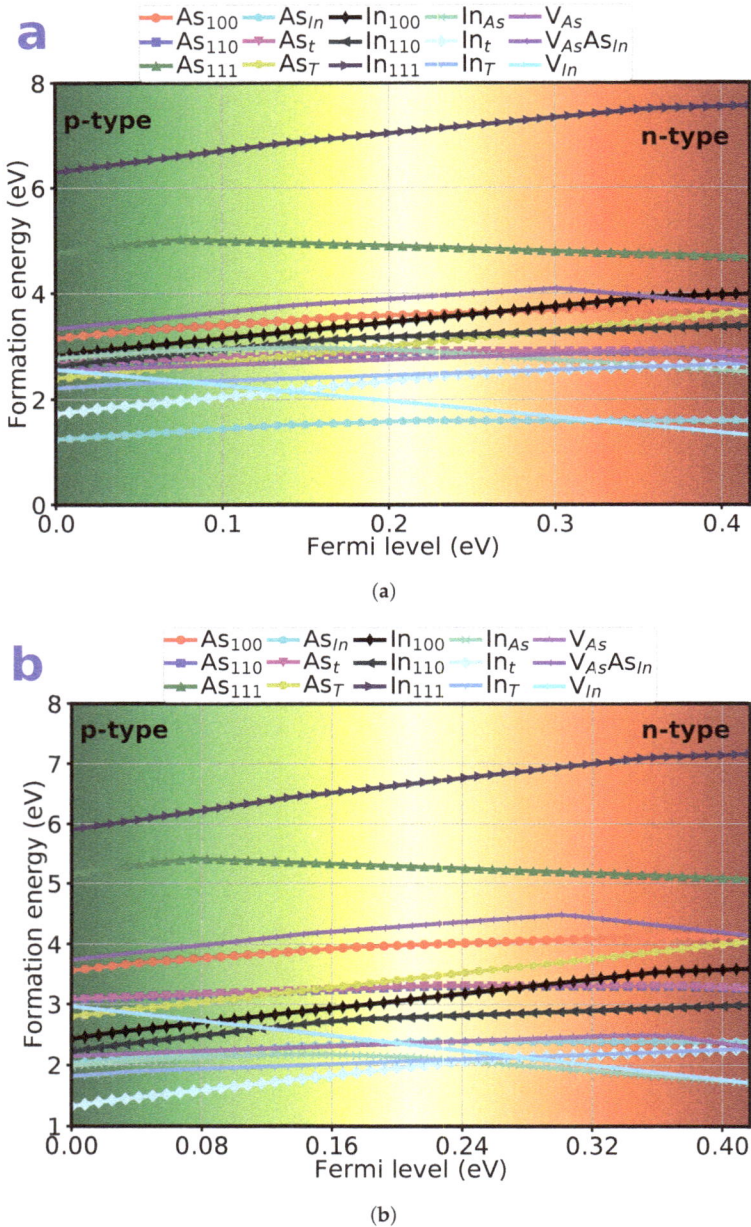

Figure 3. Formation energies as functions of E_F in the As-rich (**a**) and In-rich (**b**) environments for fifteen different types of point defects in *zb*-InAs with all possible charge states from -4 to $+4$ $+e$. The different doping regimes near valance bands (*p*-type) and conduction bands (*n*-type) are displayed.

The defect formation energy is the minimum energy for the generation of a defect. It is more practical to analyze the minimum formation energies among all the possible charge states given the fact that electrons have much higher mobilities (over three order of magnitude) than atoms. Therefore, we define the lowest formation energy of a defect as the minimum formation energy with respect to all possible charge states with $-4 \leq q \leq 4$. We would focus on the discussion of the lowest formation energies of these defects in the following subsections. For simplicity, the formation energy refers to the lowest formation energy among various charge states hereafter until specified. Such lowest formation energies as functions of E_F for each of fifteen defective configurations are presented in Figures S1 and S2 in the Supplementary Information for As-rich and In-rich environments, respectively. Here, we only discuss the lowest formation energies for five selected defect groups in the following subsections.

3.3. Formation Energy of Dumbbell Interstitials

As one atom squeezes itself into a lattice site taken by the same species in the pristine configuration, a dumbbell interstitial-type point defect will be formed. We explicitly examined six dumbbell interstitials of As_{100}, As_{110}, As_{111}, In_{100}, In_{110}, In_{111} depicted in Figure 2. The configuration of In_{111} after relaxation becomes much different from the original $<111>$ dumbbell structure (not shown here but similar to Figure 2c for As_{111} with species switched). This result indicates that In_{111} is unstable and will relax to a complex simultaneously. The formation energies of these six dumbbell interstitials as function of E_F in As-rich and In-rich chemical environments are shown in the left and right panel, respectively, of Figure 4. In_{111} and As_{111} are found to have the two highest formation energies for both cases, indicating that the $<111>$ dumbbell configurations are unfavorable for either environment and both In and As atoms.

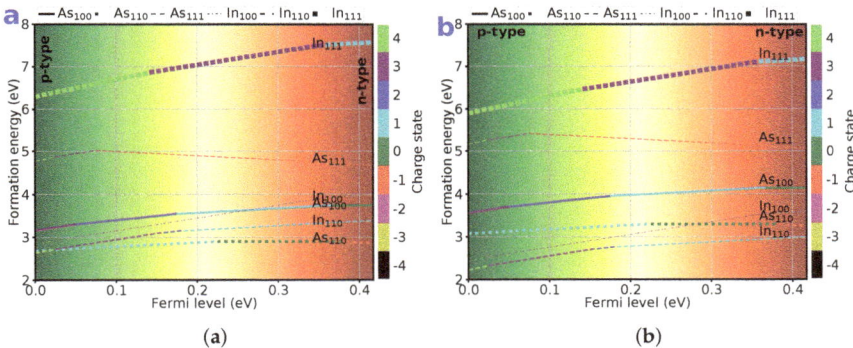

Figure 4. Formation energies as functions of E_F under (**a**) As-rich and (**b**) In-rich environment for six labeled dumbbell type point defects in *zb*-InAs with all possible charge states from -4 to $+4$ marked with different colors.

For an As-rich environment, the As_{110} has over all the lowest energy among the six dumbbell interstitials, followed by In_{110} and In_{100}. It is worth noting that in the very light *p*-doped region $E_F < 0.02$ eV, In_{110} has slightly (<0.03 eV) lower formation energy than that of As_{110}, which should be indistinguishable within this DFT study. For an In-rich environment, the configurations with the lowest three formation energies are In_{110}, In_{100}, and As_{110}, similar to the As-rich case with switched elements. Therefore, we can conclude that the dumbbell interstitials prefer $<110>$ configurations in *zb*-InAs under various charge states, chemical environment, and Fermi levels.

3.4. Formation Energy of Tetrahedral Interstitials

The tetrahedral sites are among the most energetic favorable sites for interstitials in diamond and zinc-blende structures. The four intrinsic tetrahedral interstitials are studied explicitly for (1) As on 4-As formed tetrahedral site As_t, (2) As on 4-In formed tetrahedral site As_T, (3) In on 4-As formed tetrahedral site In_t, (4) In on 4-In formed tetrahedral site In_T. Their corresponding formation energies are displayed in Figure 5 as functions of E_F for both As-rich and In-rich chemical environments. In the whole Fermi level, In_t has the lowest formation energy for both chemical environments, followed by In_T. When $E_F > 0.26$ eV, In_T has the same formation energy as In_t structure. In the very light p-doped region of $E_F < 0.02$ eV, In_t prefers $+4e$ charge states. With the increase of doping, In_t prefers $+3e$ charge state in the p-doped region but $+1e$ in the n-type doped region. The As tetrahedral interstitials are much less common in all chemical environments because their formation energy are much larger than that of In tetrahedral interstitials. As_T prefers $+3e$ charge state, opposed to As_t with $+1e$ or neutral state, in both chemical environments. The energy difference between As and In tetrahedral interstitials is larger in In-rich environment then that in As-rich environment.

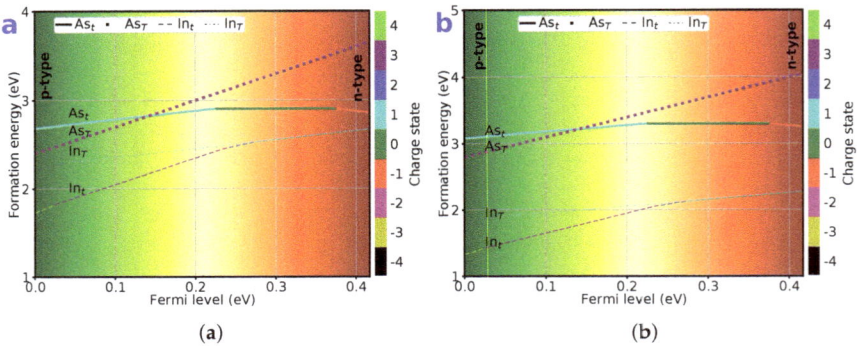

Figure 5. Formation energies as functions of E_F in the (**a**) As-rich and (**b**) In-rich environment for four tetrahedral-site interstitial type point defects in *zb*-InAs with all possible charge states from -4 to $+4$ marked with different colors. The y-axises are enlarged for viewing details.

The most interesting feature of the formation energy of tetrahedra interstitials is that for both As-rich and In-rich environments, In atoms prefer the 4-As tetrahedral site until $E_F > 0.26$ eV, where In atoms have the same formation energy at both tetrahedral sites with the same charge state. The identical formation energies and charge states suggest a fast diffusion for In atoms along the $t - T - t - T$ path among tetrahedral sites. Additionally, in an In-rich environment, the formation energies of In tetrahedral interstitials are very low, less than 2 eV. Furthermore, for the p-doping case, In_t has a small formation energy of 1.3 eV and with the charge state of $q = +3e$. Here, the low defect formation energy implies that the concentration of the corresponding defects is high under thermal equilibrium. Therefore, in the In-rich environment, In atoms prefer a tetrahedral sites with fast diffusion along the $t - T - t$ path. The quantities that describe the diffusion dynamics, including diffusion energy barrier and diffusion coefficients, deserve further study.

Finally, in In-rich environment, the formation energies of both As_T and As_t are much higher (about 2 eV) than those of In counterparts, implying that As tetrahedral interstitials are energetically unlikely to form. As a contrast, in the As-rich environment, all the formation energies of As and In interstitials become close to each other. As a result, As-type tetrahedral interstitials are preferred in this case with lower formation energies.

3.5. Formation Energy of Substitutionals

Next, we consider substitutional intrinsic point defects. Since we have two elements in the pristine *zb*-InAs , there are only two substitutional intrinsic point defects, i.e., As_{In}, and In_{As}. The formation energies of these two substitutionals as functions of E_F are displayed in Figure 6 for both As-rich and In-rich chemical environments. As seen in this figure, the formation energies of substitutionals are greatly affected by chemical environment. Under an As-rich environment, As_{In} has a relatively low formation energy, and is at least 1 eV lower than that of In_{As}. This result indicates that As atoms tend to replace In atoms at In lattice sites. As a result, the As concentration will be much higher. In addition, As substitutionals favor neutral ($q = 0$) or lower positively charged states ($q = +1, +2e$).

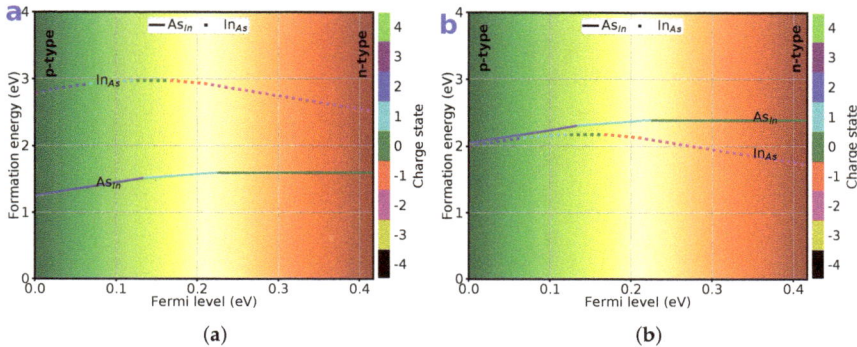

Figure 6. Formation energies as functions of E_F in the (**a**) As-rich and (**b**) In-rich environment for two substitutional type point defects in zb-InAs with all possible charge states from -4 to $+4$ marked with different colors.

On the other hand, for an In-rich environment, In substitutionals are energetically favored, with its formation energy only slightly lower than that of As substitutionals. Here, the difference of formation energy between two substitutionals is much less than that in As-rich environment, implying that the difference of concentrations between substitutionals will be much less than that in As-rich environment.

3.6. Formation Energy of Vacancies

Vacancy-type defects are known as the most common defects and play an important role in vacancy-mediated diffusion and mass transports. There are two intrinsic vacancy-type point defects in zb-InAs : V_{As} and V_{In}. We will only consider a point defect complex $V_{As}As_{In}$. This complex is of interest because it is closely related to V_{In}. When one vacancy is generated on an In atom site, one As atom on a nearest-neighbor site might dissociate from the host and fill up this vacancy site, forming the $V_{As}As_{In}$ complex. Here, we will not consider the As counterpart of $V_{In}In_{As}$, since a previous study has already reported that the defect complex of $V_{In}In_{As}$ is unstable and spontaneously relaxes back to V_{As} single vacancy-type defect [26].

The formation energies of three vacancies related to intrinsic defects are plotted in Figure 7 as functions of E_F. For both As-rich and In-rich chemical environments, V_{In} decreases quickly and monotonically with increasing E_F. In addition, the charge state prefers $q = -3e$ throughout the whole range of E_F. The decrease amounts are 1.2 and 1.3 eV for As-rich and In-rich environments, respectively. Both V_{As} and $V_{As}As_{In}$ have a general trend of increasing in formation energy with E_F up to $E_F > 0.31$ eV. The defect complex $V_{As}As_{In}$ has the highest formation energy among the three vacancy defects, larger than 3 eV, indicating that this defect complex is much less favored than two single-vacancy defects.

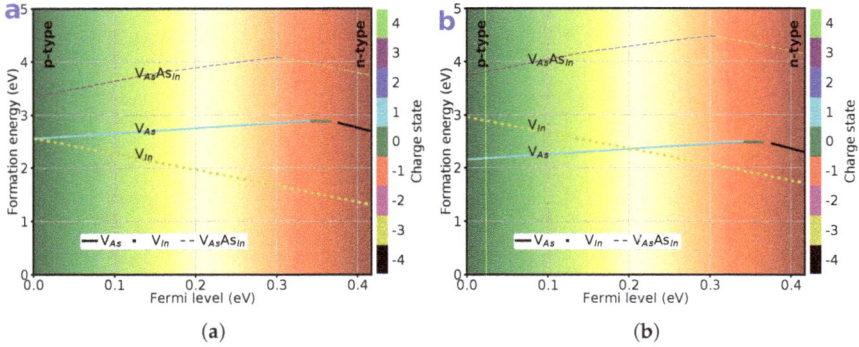

Figure 7. Formation energies as functions of E_F in the (**a**) As-rich and (**b**) In-rich environment for two vacancy-type point defects in zb-InAs with all possible charge states from -4 to $+4$ marked with different colors.

For the As-rich environment, V_{In} has the lowest formation energy among three types of vacancy-related defects. The monotonic decrease of formation energy from 2.56 eV at $E_F = 0$ to 1.35 eV at $E_F = 0.417$ eV indicates that V_{In} defects are energetically favored, especially for n-type doping. For the In-rich environment, on the other hand, V_{As} has the lowest formation energy, till $E_F > 0.2$ eV, implying that As vacancy becomes the dominant vacancy in p-doped In-rich zb-InAs . For n-type doped zb-InAs , In vacancy is the majority vacancy for all chemical environments.

3.7. Lowest 6 Formation Energy of Point Defects

After comparison of the formation energies within each individual defect group, it is insightful to compare all fifteen intrinsic point defects as a whole. For that purpose, we plotted six lowest formation energies of fifteen defect configurations in a dilute limit as functions of E_F in Figure 8 under two extreme chemical environments, i.e., As-rich and In-rich one. The six lowest formation energies ordered from low to high belongs to As_{In}, V_{In}, In_t, In_T, V_{As}, and As_{110} at $E_F = E_g/2 = 0.209$ eV at As-rich environment. The corresponding order is In_t, In_T, In_{As}, As_{In}, V_{As}, and V_{In} at In-rich environment.

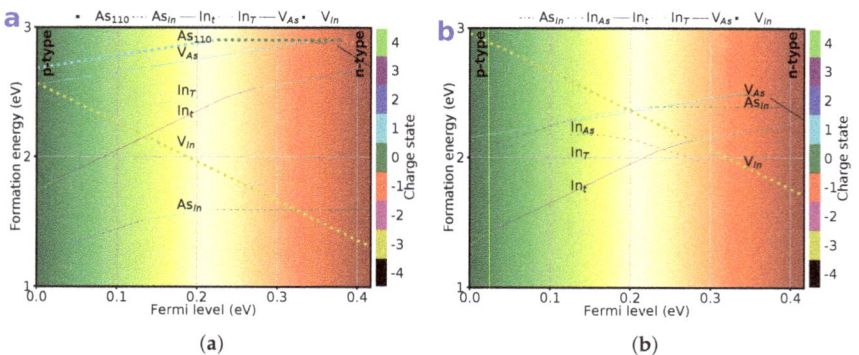

Figure 8. Six lowest defect formation energies in zb-InAs as functions of E_F in the (**a**) As-rich and (**b**) In-rich environment with all possible charge states from -4 to $+4$ marked by corresponding colors.

The chemical environment greatly changes the formation energies of intrinsic point defects. Six lowest formation energies in an As-rich environment differ from those in In-rich environment for species, amount, and charge states. For As-rich environments, the lowest formation energy is 1.24 eV for As$_{In}$ at $E_F = 0$ win the charge state of $q = +2e$, and 1.59 eV for $E_F > 0.22$ eV in the charge-neutral state. Our results qualitatively agree with an earlier report that the As$_{In}$ defect has the lowest formation energy in arsenic rich environments with the same charge status but much less formation energy. [55] The discrepancy might attribute to the small supercell (only 64 atoms) in their study. The second lowest formation energy belongs to vacancy-type V$_{In}$ with the formation energy as low as 1.31 eV at $E_F = 0.417$ eV. All the other point defects have their formation energies larger than 1.6 eV, indicating much smaller concentrations than those of As$_{In}$ and V$_{In}$ defects.

On the other hand, for an In-rich environment, the lowest formation energies are among In tetrahedral interstitials and substitutionals. For *p*-type doping, In$_t$ interstitials are dominant with the lowest formation energy for $E_F < 0.24$ eV. Actually, its smallest value is 1.33 eV at $E_F = 0$. When $E_F > 0.24$ eV, the lowest formation energy among all intrinsic point defects is switched to substitutional In$_{As}$ with the minimum value of 1.715 eV at $E_F = 0.417$ eV and the charge state of $q = -2e$. It is worth noting that a previous study reported that In$_t$ interstitials have lowest formation energy, ranged from 0.3 to 1.55 eV, through out the whole Fermi level. [55] The difference might be due to the much smaller system size in the previous study.

Overall, the minimum formation energy of an intrinsic point defect is predicted to be 1.24 eV in *zb*-InAs with As$_{In}$ configuration. This is a substitutional point defect in an As-rich chemical environment with the charge state of $q = +2e$ in light doping regime. The predominant substitutional point defects are expected useful in designing radiation-tolerant electronics and opto-electronics, because the ubiquitous point defects reduce the separation of Frenkel pairs, and therefore enhance the recombination of point defects under irradiation of high-energy particles. To increase the radiation resistance, it is suggested to choose an As-rich chemical environment.

As a side note, these point defects accessed via first-principles calculations here could be detected experimentally by photo-luminescence (PL) or Electron paramagnetic resonance (EPR) spectroscopy [21]. Further experimental investigations are desirable for their irradiation tolerance study.

4. Conclusions

We have systematically investigated fifteen different kinds of intrinsic point defects in single crystalline *zb*-InAs using a supercell method by means of first-principles calculations within the frame of density functional theory. These fifteen types of intrinsic point defects have been characterized as five groups, namely dumbbell interstitials, tetrahedral interstitials, substitutionals, vacancies, and complex. We have examined the formation energies of all of these intrinsic point defects as functions of Fermi level E_F in both As-rich and In-rich chemical environments with charge states ranging from $-4e$ to $+4e$ in the dilute-solution limit including finite-size corrections. All fifteen types of defect formation energies are found greatly affected by chemical environments. For As-rich environment, substitutional point defects are the primary intrinsic point defects in *zb*-InAs until the *n*-type doped region $E_F > 0.32$ eV is reached, where the dominant intrinsic point defects are changed to In vacancies. For In-rich environment, In tetrahedral interstitial has the lowest formation energy till *n*-type doped $E_F > 0.24$ eV region where substitutional point defects In$_{As}$ take over. The $< 111 >$ dumbbell interstitials are found unfavorable among all the point defects. However, they prefer $< 110 >$ configurations, instead. The most interesting feature of the tetrahedral interstitials is that for both As-rich and In-rich environments, In atoms prefer the 4-As tetrahedral site In$_t$ up to $E_F > 0.26$ eV, where In atoms acquire the same formation energy at both tetrahedral sites and the same charge state. The identical formation energies and charge states strongly suggest a fast diffusion process for In atoms along the path of $t - T - t$ among various tetrahedral sites. The In-rich chemical environment greatly reduces the formation energy of In tetrahedral sites, implying much higher concentrations. In addition, V$_{In}$ decreases quickly and monotonically with increasing E_F, and its charge state prefers $q = -3e$ throughout the whole range of E_F. The most popular vacancy-type defect is V$_{In}$ in an As-rich environment, but switches to V$_{As}$ in an In-rich environment at *p*-type doped region of $E_F < 0.2$ eV.

Our results shed light on relative stabilities of these intrinsic point defects, as well as their relative concentrations and possible diffusions. This study is expected very insightful in defect-engineering the *zb*-InAs based semiconductors, as well as material design for radiation resistant electronics and opto-electronics.

Supplementary Materials: The following are available online at http://www.mdpi.com/2073-4352/9/1/48/s1: Figure S1: Defect formation energy in As-rich environment, Figure S2: Defect formation energy in In-rich environment.

Author Contributions: Conceptualization, F.G., D.H. and E.R.H.; computations, Q.P. and N.C.; writing-original draft preparation, Q.P.; writing-review and editing, F.G., D.H., E.R.H., D.A.C., N.C.; supervision, F.G. and D.H.; project administration, F.G.

Funding: This research was funded by Air Force Research Laboratory (AFRL) and Air Force Office of Scientific Research (AFOSR)

Acknowledgments: The authors would like to acknowledge the generous financial support from the Air Force Research Laboratory (AFRL). D.H. would like to thank the Air Force Office of Scientific Research (AFOSR) for support.

Conflicts of Interest: There are no conflicts of interest to declare.

Data Availability: The datasets generated during and/or analyzed during the current study are available from the corresponding author on reasonable request.

References

1. Razeghi, M. High-power laser-diodes based on ingaasp alloys. *Nature* **1994**, *369*, 631–633. [CrossRef]
2. Kim, S.-W.; Sujith, S.; Lee, B.Y. InAsxSb1-x alloy nanocrystals for use in the near infrared. *Chem. Commun.* **2006**, 4811–4813. [CrossRef]
3. Marshall, A.R.J.; Tan, C.H.; Steer, M.J.; David, J.P.R. Electron dominated impact ionization and avalanche gain characteristics in InAs photodiodes. *Appl. Phys. Lett.* **2008**, *93*, 111107. [CrossRef]
4. Chelikowsky, J.R.; Cohen, M.L. Nonlocal pseudopotential calculations for the electronic structure of eleven diamond and zinc-blende semiconductors. *Phys. Rev. B* **1976**, *14*, 556–582. [CrossRef]
5. Gorczyca, I.; Christensen, N.E.; Alouani, M. Calculated optical and structural properties of InP under pressure. *Phys. Rev. B* **1989**, *39*, 7705–7712. [CrossRef]
6. Chang, K.J.; Froyen, S.; Cohen, M.L. Pressure coefficients of band gaps in semiconductor. *Solid State Commun.* **1984**, *50*, 105–107. [CrossRef]
7. Del Alamo, J.A. Nanometre-scale electronics with III-V compound semiconductors. *Nature* **2011**, *479*, 317–323. [CrossRef]
8. Jones, K.S.; Lind, A.G.; Hatem, C.; Moffatt, S.; Ridgway, M.C. A Brief Review of Doping Issues in III-V Semiconductors. *ECS Trans.* **2013**, *53*, 97–105. [CrossRef]
9. Dayeh, S.A.; Susac, D.A.; Kavanagh, K.L.; Yu, E.T.; Wang, D. Structural and Room-Temperature Transport Properties of Zinc Blende and Wurtzite InAs Nanowires. *Adv. Funct. Mater.* **2009**, *19*, 2102–2108. [CrossRef]
10. Adachi, S. *Properties of Semiconductor Alloys: Group-IV, III-V and II-VI Semiconductors*; Wiley Online Library: Hoboken, NJ, USA, 2009.
11. Park, H.D.; Prokes, S.M.; Cammarata, R.C. Growth of epitaxial InAs nanowires in a simple closed system. *Appl. Phys. Lett.* **2005**, *87*, 063110. [CrossRef]
12. Dayeh, S.A.; Yu, E.T.; Wang, D. III-V nanowire growth mechanism: V/III ratio and temperature effects. *Nano Lett.* **2007**, *7*, 2486–2490. [CrossRef] [PubMed]
13. Dick, K.A.; Deppert, K.; Martensson, T.; Mandl, B.; Samuelson, L.; Seifert, W. Failure of the vapor-liquid-solid mechanism in Au-assisted MOVPE growth of InAs nanowires. *Nano Lett.* **2005**, *5*, 761–764. [CrossRef] [PubMed]
14. Bashouti, M.Y.; Tung, R.T.; Haick, H. Tuning the Electrical Properties of Si Nanowire Field-Effect Transistors by Molecular Engineering. *Small* **2009**, *5*, 2761–2769. [CrossRef]
15. Dayeh, S.A.; Aplin, D.P.R.; Zhou, X.; Yu, P.K.L.; Yu, E.T.; Wang, D. High electron mobility InAs nanowire field-effect transistors. *Small* **2007**, *3*, 326–332. [CrossRef] [PubMed]

16. Dayeh, S.A.; Yu, E.T.; Wang, D. Transport Coefficients of InAs Nanowires as a Function of Diameter. *Small* **2009**, *5*, 77–81. [CrossRef] [PubMed]

17. Park, D.W.; Jeon, S.G.; Lee, C.-R.; Lee, S.J.; Song, J.Y.; Kim, J.O.; Noh, S.K.; Leem, J.-Y.; Kim, J.S. Structural and electrical properties of catalyst-free Si-doped InAs nanowires formed on Si(111). *Sci. Rep.* **2015**, *5*, 16652. [CrossRef] [PubMed]

18. Ning, F.; Tang, L.-M.; Zhang, Y.; Chen, K.-Q. First-principles study of quantum confinement and surface effects on the electronic properties of InAs nanowires. *J. Appl. Phys.* **2013**, *114*, 224304. [CrossRef]

19. Alam, K.; Sajjad, R.N. Electronic Properties and Orientation-Dependent Performance of InAs Nanowire Transistors. *IEEE Trans. Electron Devices* **2010**, *57*, 2880–2885. [CrossRef]

20. Dos Santos, C.L.; Piquini, P. Diameter dependence of mechanical, electronic, and structural properties of InAs and InP nanowires: A first-principles study. *Phys. Rev. B* **2010**, *81*, 075408. [CrossRef]

21. Repp, S.; Weber, S.; Erdem, E. Defect Evolution of Nonstoichiometric ZnO Quantum Dots. *J. Phys. Chem. C* **2016**, *120*, 25124–25130. [CrossRef]

22. Dodd, P.E.; Massengill, L.W. Basic mechanisms and modeling of single-event upset in digital microelectronics. *IEEE Trans. Nucl. Sci.* **2003**, *50*, 583–602. [CrossRef]

23. Ziegler, J.F.; Lanford, W.A. Effect of Cosmic Rays on Computer Memories. *Science* **1979**, *206*, 776–788. [CrossRef] [PubMed]

24. Dick, K.A.; Caroff, P.; Bolinsson, J.; Messing, M.E.; Johansson, J.; Deppert, K.; Wallenberg, L.R.; Samuelson, L. Control of III-V nanowire crystal structure by growth parameter tuning. *Semicond. Sci. Technol.* **2010**, *25*, 024009. [CrossRef]

25. Kim, Y.-S.; Hummer, K.; Kresse, G. Accurate band structures and effective masses for InP, InAs, and InSb using hybrid functionals. *Phys. Rev. B* **2009**, *80*, 035203. [CrossRef]

26. Reveil, M.; Huang, H.-L.; Chen, H.-T.; Liu, J.; Thompson, M.O.; Clancy, P. Ab Initio Studies of the Diffusion of Intrinsic Defects and Silicon Dopants in Bulk InAs. *Langmuir* **2017**, *33*, 11484–11489. [CrossRef] [PubMed]

27. Wang, J.; Lukose, B.; Thompson, M.O.; Clancy, P. Ab initio modeling of vacancies, antisites, and Si dopants in ordered InGaAs. *J. Appl. Phys.* **2017**, *121*, 045106. [CrossRef]

28. Northrup, J.E.; Zhang, S.B. Dopant and defect energetics - Si in GaAs. *Phys. Rev. B* **1993**, *47*, 6791–6794. [CrossRef]

29. Lee, C.-W.; Lukose, B.; Thompson, M.O.; Clancy, P. Energetics of neutral Si dopants in InGaAs: An ab initio and semiempirical Tersoff model study. *Phys. Rev. B* **2015**, *91*, 094108. [CrossRef]

30. Lee, S.G.; Chang, K.J. Energetics and hydrogen passivation of carbon-related defects in InAs and In0.5Ga0.5As. *Phys. Rev. B* **1996**, *53*, 9784–9790. [CrossRef]

31. Freysoldt, C.; Grabowski, B.; Hickel, T.; Neugebauer, J.; Kresse, G.; Janotti, A.; Van de Walle, C.G. First-principles calculations for point defects in solids. *Rev. Mod. Phys.* **2014**, *86*, 253–305. [CrossRef]

32. Leslie, M.; Gillan, N.J. The energy and elastic dipole tensor of defects in ionic crystals calculated by the supercell method. *J. Phys. C Solid State Phys.* **1985**, *18*, 973. [CrossRef]

33. Schultz, P.A. Charged Local Defects in Extended Systems. *Phys. Rev. Lett.* **2000**, *84*, 1942–1945. [CrossRef] [PubMed]

34. Makov, G.; Payne, M.C. Periodic boundary conditions in ab initio calculations. *Phys. Rev. B* **1995**, *51*, 4014–4022. [CrossRef]

35. Van de Walle, C.G.; Neugebauer, J. First-principles calculations for defects and impurities: Applications to III-nitrides. *J. Appl. Phys.* **2004**, *95*, 3851–3879. [CrossRef]

36. Freysoldt, C.; Neugebauer, J.; Van de Walle, C.G. Fully Ab Initio Finite-Size Corrections for Charged-Defect Supercell Calculations. *Phys. Rev. Lett.* **2009**, *102*, 016402. [CrossRef] [PubMed]

37. Lany, S.; Zunger, A. Assessment of correction methods for the band-gap problem and for finite-size effects in supercell defect calculations: Case studies for ZnO and GaAs. *Phys. Rev. B* **2008**, *78*, 235104. [CrossRef]

38. Komsa, H.-P.; Rantala, T.T.; Pasquarello, A. Finite-size supercell correction schemes for charged defect calculations. *Phys. Rev. B* **2012**, *86*, 045112. [CrossRef]

39. Castleton, C.W.M.; Höglund, A.; Mirbt, S. Managing the supercell approximation for charged defects in semiconductors: Finite-size scaling, charge correction factors, the band-gap problem, and the ab initio dielectric constant. *Phys. Rev. B* **2006**, *73*, 035215. [CrossRef]

40. Taylor, S.E.; Bruneval, F. Understanding and correcting the spurious interactions in charged supercells. *Phys. Rev. B* **2011**, *84*, 075155. [CrossRef]

41. Kresse, G.; Hafner, J. Ab initio molecular dynamics for liquid metals. *Phys. Rev. B* **1993**, *47*, 558–561. [CrossRef]

42. Kresse, G.; Furthuller, J. Efficiency of ab-initio total energy calculations for metals and semiconductors using a plane-wave basis set. *Comput. Mater. Sci.* **1996**, *6*, 15–50. [CrossRef]

43. Hohenberg, P.; Kohn, W. Inhomogeneous electron gas. *Phys. Rev.* **1964**, *136*, B864. [CrossRef]

44. Kohn, W.; Sham, L.J. Self-consistent equations including exchange and correlation effects. *Phys. Rev.* **1965**, *140*, A1133. [CrossRef]

45. Perdew, J.P.; Burke, K.; Ernzerhof, M. Generalized Gradient Approximation Made Simple. *Phys. Rev. Lett.* **1996**, *77*, 3865–3868. [CrossRef] [PubMed]

46. Perdew, J.P.; Ruzsinszky, A.; Csonka, G.I.; Vydrov, O.A.; Scuseria, G.E.; Constantin, L.A.; Zhou, X.; Burke, K. Restoring the density-gradient expansion for exchange in solids and surfaces. *Phys. Rev. Lett.* **2008**, *100*, 136406. [CrossRef]

47. Blöchl, P.E. Projector augmented-wave method. *Phys. Rev. B* **1994**, *50*, 17953–17979. [CrossRef]

48. Jones, R.O.; Gunnarsson, O. The density functional formalism, its applications and prospects. *Rev. Mod. Phys.* **1989**, *61*, 689–746. [CrossRef]

49. Bechstedt, F.; Belabbes, A. Structure, energetics, and electronic states of III-V compound polytypes. *J. Phys. Condens. Matter* **2013**, *25*, 273201. [CrossRef] [PubMed]

50. Tahini, H.A.; Chroneos, A.; Murphy, S.T.; Schwingenschlögl, U.; Grimes, R.W. Vacancies and defect levels in III–V semiconductors. *J. Appl. Phys.* **2013**, *114*, 063517. [CrossRef]

51. Madelung, O.; Schulz, M.; Weiss, H. (Eds.) *Semiconductors, Physics of Group IV Elements and III-V Compounds*; Springer-Verlag: New York, NY, USA, 1982; Volumn 17.

52. Shu, H.; Liang, P.; Wang, L.; Chen, X.; Lu, W. Tailoring electronic properties of InAs nanowires by surface functionalization. *J. Appl. Phys.* **2011**, *110*, 103713. [CrossRef]

53. Thompson, A.G.; Rowe, J.E.; Rubenste, M. Preparation and Optical Properties of InAs1-xPx Alloys. *J. Appl. Phys.* **1969**, *40*, 3280–3288. [CrossRef]

54. Dreyer, C.E.; Alkauskas, A.; Lyons, J.L.; Janotti, A.; Van de Walle, C.G. First-Principles Calculations of Point Defects for Quantum Technologies. *Annu. Rev. Mater. Res.* **2018**, *48*, 1–26. [CrossRef]

55. Höglund, A.; Castleton, C.W.M.; Göthelid, M.; Johansson, B.; Mirbt, S. Point defects on the (110) surfaces of InP, InAs, and InSb: A comparison with bulk. *Phys. Rev. B* **2006**, *74*, 075332. [CrossRef]

MDPI

St. Alban-Anlage 66

4052 Basel

Switzerland

Tel. +41 61 683 77 34

Fax +41 61 302 89 18

www.mdpi.com

Crystals Editorial Office

E-mail: crystals@mdpi.com

www.mdpi.com/journal/crystals

www.ingramcontent.com/pod-product-compliance
Lightning Source LLC
Chambersburg PA
CBHW051912210326
41597CB00033B/6118